Residential Real Estate

Residential Real Estate introduces readers to the economic fundamentals and emerging issues in housing markets. The book investigates housing market issues within local, regional, national and international contexts in order to provide students with an understanding of the economic principles that underpin residential property markets. Key topics covered include:

- Location choice in urban areas
- Housing supply and demand
- Housing finance and housing as an asset class
- Demographic shifts and implications for housing
- Sustainable homes and digitalisation in housing

Drawing on market-level information, readers are encouraged to recognise the supply and demand drivers and modelling of dynamic housing markets at various spatial scales and the implications of trends within an urban and regional context, e.g. urbanisation, ageing population, migration, digitalisation. With research-based discussions and coverage of relevant literature, this is an ideal textbook for students of residential real estate, property and related business studies courses at UG and PG levels, as well as a reference book with research topics for researchers. This book will also be of interest to professionals and policymakers.

Anupam Nanda is Professor of Urban Economics and Real Estate, and Research Division Lead (Real Estate & Planning) at the Henley Business School, University of Reading, UK. Anupam grew up in India, and before joining the University of Reading, he worked with Deloitte & Touche in Mumbai and the National Association of Home Builders (NAHB) in Washington, DC. Anupam has also taught undergraduate Economics and Public Finance at the University of Connecticut, US, from where he obtained his PhD in Economics. Anupam holds professional fellowships of FRSA (Fellow of the Royal Society of Arts), FRGS (Fellow of the Royal Geographical Society), FRSS (Fellow of the Royal Statistical Society) and FHEA (Fellow of the Higher Education Academy, UK). His academic papers have been published in leading journals including *Journal of Urban Economics*, *Regional Studies*, *Energy Economics*, *Energy Policy*, *Review of International Economics*, *Real Estate Economics* and *Applied Linguistics*.

Residential Real Estate

Urban & Regional Economic Analysis

Anupam Nanda

LONDON AND NEW YORK

First published 2019
by Routledge
2 Park Square, Milton Park, Abingdon, Oxon OX14 4RN

and by Routledge
52 Vanderbilt Avenue, New York, NY 10017

Routledge is an imprint of the Taylor & Francis Group, an informa business

© 2019 Anupam Nanda

The right of Anupam Nanda to be identified as author of this work has been asserted by him in accordance with sections 77 and 78 of the Copyright, Designs and Patents Act 1988.

All rights reserved. No part of this book may be reprinted or reproduced or utilised in any form or by any electronic, mechanical, or other means, now known or hereafter invented, including photocopying and recording, or in any information storage or retrieval system, without permission in writing from the publishers.

Trademark notice: Product or corporate names may be trademarks or registered trademarks, and are used only for identification and explanation without intent to infringe.

British Library Cataloguing-in-Publication Data
A catalogue record for this book is available from the British Library

Library of Congress Cataloging-in-Publication Data
Names: Nanda, Anupam, author.
Title: Residential real estate: urban & regional economic analysis/Anupam Nanda.
Description: Abingdon, Oxon; New York, NY: Routledge, 2019.
Identifiers: LCCN 2018044530 | ISBN 9781138898301 (hardback) | ISBN 9781138898318 (pbk.) | ISBN 9781315708645 (ebook)
Subjects: LCSH: Residential real estate. | Housing.
Classification: LCC HD1390.5 .N36 2019 | DDC 333.33/8—dc23
LC record available at https://lccn.loc.gov/2018044530

ISBN: 978-1-138-89830-1 (hbk)
ISBN: 978-1-138-89831-8 (pbk)
ISBN: 978-1-315-70864-5 (ebk)

Typeset in Times
by Deanta Global Publishing Services, Chennai, India

I dedicate this book to my parents and other members of my circle of trust.

Contents

List of figures x
List of tables xii
Acknowledgements xiv

1 **Introduction** 1

2 **Attributes of housing markets** 4
   Chapter outline 4
   Territoriality or Spatiality or Immovability or spatial fixity in the housing market 4
   Variability and heterogeneity in the housing market 5
   Longevity or durability of housing units 5
   Informationally imperfect market 5
   Transaction costs in the housing market 6
   Externalities and external effects in the housing market 6
   Socio-political influences in the housing market 6
   Housing and the role of sentiment, personal traits and culture 7
   Selected research topics 15

3 **Economy, housing demand and supply** 16
   Chapter outline 16
   Contribution of housing to GDP 21
   The multiplier effect 25
   Demand for housing 25
   Violating the law of demand 26
   Comparison of housing with other real estate sectors 35
   Housing market area 36
   How do we identify sub-regional housing market areas? 37
   Housing data sources 37
   Selected research topics 38

4 **House price: Measures and estimation** 39
   Chapter outline 39
   Why do we need house price index? 40

House price index 41
Hedonic model 43
Empirical aspects of hedonic framework 44
Hedonic estimation strategy 48
Price index example 50
Repeat sales model 51
An example 51
Global house prices 57
Selected research topics 60

5 Location choice in urban areas 61
Chapter outline 61
Facts and figures of global urbanisation 61
Total population growth 62
Urbanisation: Regional comparisons 62
Urban agglomerations, or more megacities 63
Additional facts and figures about conditions in human settlements from the United Nations Centre for Human Settlements (UNCHS) (Habitat)'s the State of the World's Cities: 2001 64
Economics of land-use planning 77
Challenges for land-use planning 79
Housing and neighbourhood 80
Selected research topics 84

6 Housing tenure, search and mobility 85
Chapter outline 85
Housing tenure choices 88
Shared ownership 97
Housing mobility 105
Housing search process 109
Selected research topics 116

7 Demographic shifts and housing 117
Chapter outline 117
Migration and housing 117
A model explaining the impact of migrants on the demand for housing 135
Concerns of affordability 135
Selected research topics 162

8 Housing finance and the mortgage market 163
Chapter outline 163
Mortgage market process 166
Types of mortgage financing 168

Secondary mortgage market products   171
Sources of mortgage financing   173
Selected research topics   178

9 Residential real estate as an asset class   179
Chapter outline   179
Private renter sector   180
Student housing   183
Supply of student housing   184
Retirement housing   186
Cross-border investment in housing   186
Opportunities for investing in technology for residential real estate   188
Big data   190
Artificial intelligence   191
Machine learning   191
Neural network   191
Deep learning   192
Natural Language Processing (NLP)   192
Blockchain technology   192
Technology application areas in the housing market   193
Selected research topics   197

10 Housing within sustainable, intelligent places   198
Chapter outline   198
Key challenges faced by urban areas   199
Cities ranked on key parameters   206
Issues and challenges for creating an intelligent place   211
Issue 1: Innovation meets investment   215
Issue 2: Business model for 'intelligent place' projects   216
Issue 3: Urban big data and complex urban operating system   217
Issue 4: Intelligent place risks   220
Digital housing   220
Home as a hub; healthcare at home   221
Selected research topics   223

Bibliography   225
Index   242

Figures

2.1	Semiotic principles: Triad of semiotics	12
3.1	A simple model of the economy: Circular flow diagram	17
3.2	United Kingdom GDP	18
3.3	GDP (current US$ trillions, 2017) across the countries and regions	19
3.4	GDP (current US$ trillion, 2017) – BRICS	20
3.5	GDP (current US$ billion) – India and China	20
3.6	GDP growth (annual %) – India and China	21
3.7	Housing's contribution to US GDP	22
3.8	Household's sector value of land and its overlying assets, 1995 to 2016 (£ trillion)	24
3.9	Demand curve	26
3.10	Supply curve	27
3.11	Short-run supply curve	28
3.12	Market equilibrium	30
3.13	Market clearing adjustments	30
3.14	Indifference curves	31
3.15	Price determination	32
3.16	Market adjustments	34
4.1	Households and firms equilibrium, combining labour market dynamics	44
4.2	Household bid and product offer curves	47
4.3	Hedonic equilibrium and price function	48
4.4	UK house price: Comparison by methods	54
4.5	US house price: Comparison by methods	55
4.6	Ripple effect in the UK	56
4.7	Global house price index	57
4.8	Global house price growth	58
4.9	Global credit growth	58
4.10	Global housing price-to-income ratio	59
4.11	Global house price and rent ratios	60
5.1	Degree of Urbanisation	63
5.2	Estimated hourly labour costs for the whole economy in Euros, 2016 Enterprises with 10 or more employees	67

Figures xi

5.3	Premium office rent	68
5.4	Credit availability, 2017	68
5.5	Industrial electricity prices in the IEA International Energy Agency	69
5.6	Industrial gas prices in the IEA International Energy Agency	69
5.7	Bid rent curves	75
6.1	Dwelling stock: By tenure, global comparison	92
6.2	Dwelling stock: By tenure, England	92
6.3	Dwelling stock: By tenure, Great Britain	93
6.4	Household tenure status, England and Wales Census 2011	95
6.5	Public spending on supporting social rental housing in selected OECD countries	106
6.6	Residential relocation and drivers at different scales	108
6.7	Seller's and buyer's journey	109
7.1	Total migrant stock (in millions) at mid-year by origin and by major area, region, country or area of destination, 1990–2017	119
7.2	Accommodation of foreign-born in the UK	130
7.3	Distribution of housing tenure for US natives and non-natives (2011)	131
7.4	Distribution of housing tenure for US non-natives by region of birth (2011)	132
7.5	Distribution of housing tenure for US non-natives by country of birth (2011)	133
7.6	Housing building and household projection in England	135
7.7	Demand-supply mismatch in England	136
7.8	Estimated and projected population in millions of the UK and constituent countries, mid-2016 to mid-2041	137
7.9	Demand-supply mismatch and price movement in England	137
7.10	House price inflation and income growth (£)	138
7.11	Affordability ratio across English regions	139
7.12	Affordability ratio within a region in England	139
7.13	Generations likely to stay at parental home	143
7.14	Staying at parental home by education attainment	144
7.15	Scale and geographic spread of ageing population, 1950, 2000, 2050	146
7.16	Scale and geographic spread of ageing population by country	147
7.17	Aspect of former homes that are attractive	158
7.18	Reason for moving to smaller accommodation	161
8.1	Mortgage market process	170
8.2	Household debt per cent of annual disposable income	177
9.1	Tenancy and room type by source market	184
9.2	Availability of purpose built student housing (PBSA)	185
9.3	Student housing yield compared with other asset classes	186
10.1	Framework of smart city	214
10.2	Requirement of effective place innovation	217
10.3	Urban big data	219
10.4	Urban Operating System (UOS)	219
10.5	Care within intelligent home setting	222

Tables

2.1	Ten evaluation aspects for homeowners/householders in relation to housing choice	14
3.1	US recession dates: Business cycle reference dates since World War II	22
3.2	Housing's contribution to GDP performance (US)	23
3.3	Income/employment impacts of residential construction on the US economy	24
3.4	Total housing stock in OECD and EU countries, selected years	29
3.5	Main demand drivers in major real estate sectors	36
4.1	Global real estate transparency index, 2018	40
4.2	Example of hedonic estimation results	49
4.3	UK house price indices	53
4.4	US house price indices: A comparison	54
5.1	Dwelling stock in urban and rural areas	64
5.2	Highest marginal corporate tax rates in the world	70
5.3	Knowledge Economy Index (2012)	71
5.4	Global Connectivity Index	72
5.5	Example of hedonic estimation results: Neighbourhood effects	83
6.1	Guidelines for the user cost method	89
6.2	Dwelling stock: By tenure, global comparison	91
6.3	Household tenure status, England and Wales Census 2011	94
6.4	Affordability across England and Wales	97
6.5	Housing deprived population across the income distribution	99
6.6	Basic overview of intermediate housing (SO) schemes in England	102
6.7	Summary of social rental housing across selected countries	104
6.8	Total and social rental housing stock across selected countries	106
6.9	Real estate agency definitions in England	110
6.10	International real estate brokerage commission rate comparisons	112
7.1	Life expectancy at age 60 across the world	154
7.2	Life expectancy at age 65 (years): United Kingdom, 2000–02 to 2010–12	155
7.3	Forecasts of household numbers 65 and above by age cohort and housing tenure	156
7.4	Average number of rooms per person in Europe	159
7.5	Overcrowding rates in households across the income distribution	160
8.1	Housing loan penetration by country (2011)	165

8.2	Total outstanding residential loans to GDP ratio, 2004–2014	166
8.3	Government support to mortgage market	174
8.4	Growth rate of loans for house purchase	175
8.5	Variable-rate loan characteristics	177
10.1	Largest cities in terms of population, 2018	200
10.2	Sustainable cities 2016 – ranked 1–25 and 76–100	207
10.3	Top 100 sustainable cities 2016: People – ranked 1–25 and 76–100	208
10.4	Top 100 sustainable cities 2016: Planet – ranked 1–25 and 76–100	208
10.5	Top 100 sustainable cities 2016: Profit – ranked 1–25 and 76–100	209
10.6	The global liveability index 2018	210
10.7	The European digital city index 2016	210
10.8	Definition of smart city	214

Acknowledgements

First and foremost, this book is for the housing researchers and students. I would like to thank all the students who took housing economics classes from me for their questions and queries which have helped me immensely, and my former teachers, doctoral students, fellow researchers and co-authors whose works have informed this book's content and my understanding of the subject.

I would like to thank my near and dear ones for their love and affection, and providing so much inspiration over the years: my late father – Dr Kedar Nath Nanda; mother – Aparna Nanda; elder sisters – Anindita and Ankita; brothers-in-law – Sanjay and Asitava; nieces – Anannya and Ankona; daughter – Anushka; and new members of my circle of trust – Anshika and dear Viko.

I would also like to thank Ed Needle and other publishing professionals from Routledge for their help. I have made every effort to minimise the errors and omissions. However, some errors and omissions may still have creeped through. Regardless, I take responsibility for all errors. I would be grateful if readers kindly let me know of any errors and omissions and suggested corrections as they use this book. I promise to address those and further minimise errors and omissions in the future editions of the book.

ANUPAM NANDA
University of Reading, UK

1 Introduction

Housing is a basic human need, regardless of socio-economic status or any other factors and considerations. Everybody needs a house – a shelter to protect us from elements and any other physical danger; to raise a family; to rest after a day's work. A cluster of houses creates a community – a key building block for the society. This basic 'consumption' needs and aspects of housing have not changed since the dawn of the human civilisation. However, housing has become much more than that, through the advent of money, conceptualisation of investments and returns, formation of towns and cities with a concentration of jobs and economic activities within those, and through the creation of governments at various territorial levels and other social and financial institutions to support the system. Things have simply become more complex, convoluted and fraught with conflicts of interests, anxieties of decision-making, speculations about the future and so forth. The 'investment' need has started to interfere and encroach the space for the basic 'consumption' needs, and in some cases and times, it has even started to dominate over the consumption needs. Despite all these complications, especially since the industrial revolutions, formation of several institutions and financial markets over the last century, the basic needs are still intact and very relevant. At the same time, with the internet, online technology revolution and easy availability of information, we are facing new challenges and embracing changes and hoping for new opportunities. Huge amounts of information is being generated; with ever-increasing computing ability, we are able to analyse the information. Our understanding of 'how things work' is deepening. Driverless cars, human-carrying drones and other autonomous technologies will change the urban form, locational attributes and desirability of location. While these can bring in new possibilities, they can also disrupt some traditional ways of functioning of urban markets, which can cause much tensions, anxieties, tussles among groups and associated challenges. Indeed, artificial intelligence and machine learning applications will lead to faster decision-making, optimised resource utilisation and automation of various activities. This can potentially reduce inefficiencies in transactions, close information gaps and bring processes together across the board. However, unless governed appropriately and carefully with a singular focus on human wellbeing and societal prosperity, it can also lead to increased inequalities, weakened social cohesion and public unrest with deep divisions among communities. That would not bear well for human civilisation.

It is often commented by several experts that humanity is facing few big challenges or mega-trends: humanity is living longer (ageing population); humanity is getting more and more urbanised; humanity is on the move (massive displacement of people and migration) and humanity is facing climate change. All these mega-trends are being reflected by activities that are performed in various forms of buildings. Housing is perhaps the most significant part of that. According to some estimates from Savills, 2016 and the World Bank, the global real estate market is massive – valued at almost $217 trillion – and 75% of this is residential property or housing. We are talking about a huge asset class that touches all of humanity in various ways and forms throughout our life cycles.

The broad principles of economic science provide answers to many questions, conundrums and puzzles in the housing market. In this book, I rely heavily on tools and concepts of economic science to explain the fundamentals and analyse the emerging issues. My objective is to remind ourselves of the basics with open and objective views towards the new things that are cropping up in the market. Therefore, at times, the discussion crosses and transcends the disciplinary boundaries as I analyse issues both from urban and regional contexts. While a considerable amount of discussions, due to availability of data, concentrate around advanced countries, references to emerging economies are often made, and the issues of global relevance are highlighted wherever applicable.

This book is not of an introductory nature and is not a pure-play textbook either, and it is not the most comprehensive collection of issues in the housing market. It can serve as a text or reference book for final year undergraduate and post-graduate courses. It covers a significant ground in housing analysis with in-depth coverage of the key issues. And, it can be an effective reference companion for research studies as I provide an overview of key relevant literature, refer to works of several other scholars and highlight the research issues and methods. Some parts of the book assumes prior knowledge of economics and some mathematics and statistics.

I start with a discussion of how housing is a fundamentally unique economic good. Beyond obvious economics, what are the perspectives towards housing? And how are housing and sentiment so interlinked? Chapter 2 lays the foundation of understanding housing in urban and regional contexts. It provides a framework that can address many issues in the market.

Chapter 3 puts housing in the national economy framework and provides a description and analysis of demand and supply factors in the housing market. It shows challenges and gaps in our understanding of the market forces.

Building on the previous chapter, Chapter 4 brings demand and supply sides together to characterise market equilibrium (and disequilibrium). It provides an overview of the most important and talked about concept in the market – the house price. It shows existing models of price estimation and highlights the challenges and issues.

Chapter 5 puts housing issues more formally into urban and regional contexts, with models of urban areas and land-use planning, and it analyses the most important aspects of house price variation – locational or neighbourhood attributes.

Chapter 6 deals with housing choices – weighs various options of owning and renting. It explains the standard search process of buying and selling a house, highlighting the transaction dynamics and roles of intermediaries. An international comparison of the housing transaction process is presented.

Chapter 7 broadens out the most fundamental driver of housing – population with its size and composition. How are various stages of the human life cycle attached to housing needs? It deals with housing issues of people who are migrating and of those who are young and also those who are elderly.

Chapter 8 analyses the housing finance system. Mortgage market fundamentals, lessons from financial crises and international comparisons are made.

Chapter 9 formally and explicitly deals with housing as an investment class – from the big institutional investors to foreign investors and to new forms of investment channels.

Chapter 10 takes a critical view of technology in the housing market – how smart homes, smart or intelligent cities and the technologically and digitally enhanced urban environment would change how we live and work and implications for the housing market.

All chapters highlight how the tools and concepts can come together to demonstrate the functioning of the economic models. Several sections also highlight how the economic models can be built and an empirical estimation can be performed to address key research questions. Data is key in understanding the scale and nature of the housing market. Throughout the book, I have tried to provide a comprehensive account of real data as much as possible. The readers are encouraged to consult the sources of these data for further and deeper understanding.

I have tried to provide as much international focus as practicable across the topics. While some topics have been dealt with specific references to certain countries, the issues and implications can be common and lessons and insights can be applied. Lack of useable and publicly available information has played a role in not being able to cover some of the countries.

2 Attributes of housing markets

Chapter outline

- Characterising housing markets in general terms as an economic good
- Perspective towards housing
- Housing and sentiment
- Selected research topics

A housing market can be driven by a set of attributes that play key roles in determining the demand and supply, the volume of investments, the level of consumption, the profile of the housing stock and consequently, the house prices. According to the housing literature, there are several attributes that can be identified in any housing market[1]: (a) Territoriality or Spatiality or Immovability or spatial fixity in the housing market; (b) Variability and Heterogeneity in the housing market; (c) Longevity or Durability of housing units; (d) Informationally imperfect market; (e) Transaction costs in the housing market; (f) Externality and external effects in the housing market; and (g) Socio-political influences in the housing market. Few other attributes have also been noted in the literature – the importance of financial markets, price volatility, lumpiness, housing wealth and tax treatment (Meen, 2001; Miles, 1994).

Territoriality or Spatiality or Immovability or spatial fixity in the housing market

Houses are spatially fixed once developed. Houses do not move, we move to the houses! This implies the requirement of strategic planning at the outset. In the absence of strategic spatial plans, the development becomes haphazard and public interests get compromised. This immovability or immobility attribute in a housing market makes places, and thus makes the housing demand, concentrated, and therefore requires planning for public services and amenities, raising issues of community formation and targeted housing supply. The urban forms of all cities are especially spatially concentrated due to the presence of key socio-economic activities. The expansion, increasing population and need for better and more public services are expected to make these issues even more important as land value changes across various locations.

---

1 For a discussion on some of these attributes, see Meen (2001), Miles (1994). Some of these discussions are adapted from Kavarnou and Nanda (2015).

Variability and heterogeneity in the housing market

Heterogeneity of a housing market implies variation in property types, locations, neighbourhoods, participants, public amenities, etc. Housing markets across the world portray a wide range of features that make analysis specific to the market. A common, unified framework is difficult to achieve. The heterogeneity stems from both the physical and locational characteristics. While the physical attributes are often observed, the locational attributes pose significant challenges as unobserved characteristics play significant roles. Even in smaller cities, although the market size is small, there can be significant heterogeneities across various zones and sub-zones. At the property level, structural characteristics such as size or age of a property, quality of the construction, views, architectural style of specific groups or individual islands and other facilities are only some of the property level differentiators. At the neighbourhood level, there are variations in public amenities and other unique features. In most developing countries, due to economic reforms over the last two decades, there are clear divides between the new and old housing stock. Moreover, due to public infrastructures and natural geographic features, different pockets of housing developments portray marked variation. However, within the clusters of high-rise housing developments, the stock is more homogeneous. This provides a benchmark for valuation of the location and pricing of the properties.

Longevity or durability of housing units

A key feature of the housing market is the longevity of the buildings. Houses get consumed and invested by various occupants (owners and renters) and other stakeholders over and over again during its life cycle. Not just in the physical sense, but also in terms of consumption and investment, the time horizon is long in the housing market. A house is typically consumed over decades and a housing investment is made over an average of 20–30 years. The durability or longevity of houses provides an interesting challenge of time-series analysis. Due to long horizon economic dynamics, houses experience significant fluctuation over the life cycle. While the age of the housing stock is old in developed countries, the housing stock in developing countries is much younger. From the investment horizon point of view, the turnover rate can be relatively frequent in cities with heavy investors' presence and the prevalence of cash transactions.

Informationally imperfect market

A house as an economic good is of a heterogeneous nature. Heterogeneity in the use of various attributes of a house raises the need for information in order to measure willingness to pay for the attribute. Moreover, a housing transaction typically involves several parties. More importantly, each party may have conflicting interests. At the same time, each party may also have different information sets – both in terms of quality and quantity of information. As a result, a successful transaction crucially depends on the extent of information mismatch to close the gap between asking price and offer price. This also leads to a need for intermediaries or estate agents. Information asymmetry raises several interesting research problems. Issues related to information asymmetry have been variously analysed in the literature (see, for example, Green, 2008; Favara and Song, 2014). Generally, the severity of information asymmetry reduces with appropriate regulatory frameworks and formalisation of the housing market.

In most developed markets, the presence of formal processes helps in alleviating information asymmetry. However, as in most developing economies, lack of regulatory support increases the importance and impact of information asymmetry. An informal transaction process and significant reliance on private information pose several challenges. This can lead to a widening of the gap between the 'registered value' and the 'transaction value' of a property, which can exacerbate information asymmetry.

Transaction costs in the housing market

Following on the previous attribute of information asymmetry, a high level of information mismatch necessitates reliance on private and proprietary information. Generally, an important provider of such private and proprietary information are estate agents or real estate brokers. However, this means the cost of the transaction increases. A sizeable literature exists dealing with various types and items of transaction costs (see, for example, MacLennan, 1982; Quigley, 2002) such as: (a) search costs, (b) legal and administrative costs, (c) adjustment costs, (d) financial costs and (e) costs of uncertainties. Moreover, it is not just the monetary costs, but also there are significant amounts of non-monetary costs involved in housing transactions such as time costs and psychological costs. These costs may lead to a longer 'inertia' in the housing transaction and also adds to a lagged adjustment process in the market. A longer adjustment process makes the short-run supply curve rigid for a longer time period, which leads to short-run price fluctuations. The presence of informal markets and unregulated practices can increase transaction costs (including both monetary and non-monetary items).

Externalities and external effects in the housing market

A house stands within a local environment. Therefore, the location plays the most important role in making a house more or less desirable. Several external effects are then capitalised into observable house prices due to a willingness to pay for various external amenities (or, disamenities) (Schwartz et al., 2006; Rossi-Hansberg et al., 2010). Typical examples of external effects include school quality, quality of infrastructure and crime rate. Effects of external environment on house values have been the most active research area in the housing literature. While there are several common factors, there are local and specific factors related to transportation, retail and other sectors.

Socio-political influences in the housing market

The political economy in a housing market revolves around a set of rules, regulations, policies and fiscal instruments (such as taxes and subsidies) created by various layers of governments and local authorities in order to control land use and achieve socially desirable goals. Across all housing markets, there are numerous examples of government interventions. All market interventions create institutions which require further legislation and governance. The literature in this regard is well established (see Ihlanfeldt and Boehm, 1987). For example, Lund (2011) defines housing policy as "processes involved in state intervention in the housing market". Several studies have focused on the analysis of housing policies from interdisciplinary perspectives. The supply side of the housing market is a lot more active in terms of housing policies. Most housing markets exhibit these features. Several laws and policies related to land use, planning regulations to control building activities, building code regulations and financing schemes are present. In some markets, there can also be a significant presence of informal housing markets, which poses a different challenge. Urbanisation is typically associated with increased developmental activities which often lead to significant, heterogeneous option values for developable land parcels. As such, it creates significant changes in land values that get distributed rather disproportionately among landowners and other stakeholders. Standard fiscal policies and instruments fail to capture how the costs of providing urban infrastructure and public services are socialised, and how the benefits are privatised. The concept of 'land value capture' is to channelise the uplift in value towards community improvements by 'taxing' some or all the land value increases due to public investments. A prime example is the metro or underground railways across many cities in the world. The value uplift due to improved connectivity can be of very significant dimensions and should be captured and invested back for public benefits. Therefore, we often see governments enacting more legislative controls on property development and land use.

The rapid expansion of the city limits in every direction across the world has provided various benefits of modern life along with a multitude of challenges due to unrestrained growth, including a dual economy with a high level of income inequality and crippled urban governance structure. Key socio-economic parameters such as basic public services, corruption control and privatisation of public assets and services have underperformed. Therefore, tools are required in order to create and deliver capacity building programmes responsive to the local government's needs. These may include reviews and assessments of the local government's capacities to take care of its own affairs and develop practical alliances with local businesses and community organisations in order to harness benefits.

Housing and the role of sentiment, personal traits and culture

The rational choice theory has been the backbone of modelling market outcomes in Economic Science for the last several decades (Muth, 1961; Sen, 1977; Mill, 1984). Standard neoclassical economic models represent consumers/agents as having self-interest and omniscience (complete information) and they are 'continuously-optimising', 'dynamically-consistent', 'expected-utility maximisers' (Lucas, 1976). The simplicity of assuming rationality for the representative agent lends itself neatly to the attractive feature of generalisability as that can lead us to a market-level understanding, inform and facilitate devising of public policies and financial instruments. However, those models and many policy prescriptions have repeatedly (and sometimes spectacularly) failed to explain several economic crises (e.g. 2008 Global Financial Crisis) and also frequently stumbled in explaining common economic phenomena displaying cognitive dissonance (e.g. influence of advertisement, marketing). As Thorstein Veblen (1857–1929), known as the father of institutional economics, has aptly put: *"the hedonistic conception of man is that of a lightning calculator of pleasures and pains who oscillates like a homogeneous globule of desire of happiness under the impulse of stimuli that shift him about the area, but leave him intact"*. Veblen instead perceived human economic decisions as the result of multiple complex cumulative factors:

> *it is the characteristic of man to do something, not simply to suffer pleasures and pains through the impact of suitable forces. He is ... a coherent structure of propensities and habits which seeks realization and expression in an unfolding activity. ... They are the products of his hereditary traits and his past experience, cumulatively wrought out under a given body of traditions conventionalities, and material circumstances; and they afford the point of departure for the next step in the process.*

Many factors play various roles in complex combinations in individuals' lives with distinctive stories and processes. Veblen further commented that the economic life history is characterised by several processes driven by environmental changes. Evolutionary economics also provides criticisms of the 'rational agent' assumption, citing the 'parental bent' (the idea that biological impulses can and do frequently override rational decision-making based on utility). Right on! Unfortunately, actionable research building on Veblen's view of the 'economic man' (Mill, 1984) or *Homo economicus* (Thaler, 2000) has been scant in the field of economic science, especially in the context of housing markets. It needs a multi-disciplinary approach and methodologies.

The idea that an economic agent is 'rational' is generally built around several rather restrictive key assumptions such as:

(a) *Perfect Information*: the simple rational choice model assumes that the individual has full or perfect information about the alternatives, i.e. the ranking between two alternatives involves no uncertainty.
(b) *Choice under Uncertainty*: the individual effectively is able to choose between lotteries, where each lottery induces a different probability distribution over outcomes. The additional assumption of independence of irrelevant alternatives then leads to expected utility theory.
(c) *Inter-temporal Choice*: when decisions affect choices (such as consumption) at different points in time, the standard method for evaluating alternatives across time involves discounting future payoffs.
(d) *Limited Cognitive Ability*: identifying and weighing each alternative against every other may take time, effort and mental capacity. Recognising the cost that these impose or cognitive limitations of individuals gives rise to theories of bounded rationality. These assumptions are frequently and severely violated in real world events and phenomena and daily economic behaviour, especially in the housing market context due to the special attributes we discussed earlier in the chapter.

Rational economic agents are supposed to use market information to form their individual expectations. Such expectations should notionally shape the agents' actual behaviour in the marketplace. The theoretical premise behind the importance of market sentiment in asset pricing in economics is that consumers and other market agents form their perceptions based on available information and also on expectations that cannot be explained by fundamental economic drivers. However, the agents are likely to behave according to their perceptions, attitude and sentiment. Possible explanations include the presence of '*animal spirits*' and habit persistence (Acemoglu and Scott, 1994; Akerlof and Shiller, 2009). *'Animal spirits' can be determined by a complex 'cocktail' of personal attributes such as attitude, perception, culture, value, judgement, etc.* Especially in the housing market, *'animal spirits'* can play very significant roles in choice and decision-making processes. However, empirical findings around such relationships are very mixed in the literature. This is due to a lack of personal level granular data that can potentially *un-mask* and identify the sentiment elements. It is important to note that pure sentiment is the part of the effect that does not directly depend on the economic fundamentals and available information. Therefore, the empirical strategy needs to untangle the unobserved heterogeneity of factors by separating out the 'pure' sentiment element. Therefore, understanding the psychology behind economic decision-making is the *sine qua non* for any economic analysis, and an appropriate research design is needed that can lead to paradigm-shifting outcomes and an interdisciplinary understanding which can stand the test of time and space/geography. A large number of studies in the last couple of decades made numerous attempts to understand what roles sentiment of economic agents play in shaping up consumption and investment decisions and, subsequently, specific market movements through collective or individual actions. Sentiment clearly plays a key role in economic performances, but its impact under various economic environments is not straightforward as it may change for different types of agents at different times of the economic cycle and over different geographies. Moreover, causality effects are not clear as market economies may be both driven and shaped by the agents' behaviour. While there is no doubt about the existence of this effect, its extent is rather unclear despite several attempts to estimate it using different tools and models (Marcato and Nanda, 2016).

Moreover, the issue needs to be explored further. There may be a non-perfectly symmetric response to market sentiment from economic agents taking opposite sides in a transaction. Such asymmetry may be exacerbated with new information when publicly available information is limited for the financial assets which are not traded frequently such as property/housing. This would inform a range of market players who are keen on understanding the behavioural dynamics of consumption and investment.

Housing can be seen as a conspicuous economic good that involves complex decisions that all human beings need to make at least a few times in their lives. Therefore, housing serves a multitude of purposes in terms of consumption and investments. However, public policies shaping housing markets, supply, etc., are often designed around financial considerations only. More importantly, housing evokes human emotions in which culture and perceptions play a big role but these 'soft' considerations are not directly observable and evident in aggregate housing economic and market data even though they tend to correlate with observable, quantified housing attributes. Therefore, there are several pertinent questions that can be raised in the housing market context:

- *How do personal traits and perceptions shape economic decision-making in the housing market?*
- *How can these attributes be incorporated into traditional economic models of house price determination to understand the role and significance of human emotional responses?*
- *How can such understanding inform policy formulation to tackle the complex sources and extent of housing market fluctuations?*

The answers to these questions can greatly facilitate effective policy formulation and the design of financial and policy instruments that can address sources, extent and implications of the housing market instability. The theoretical starting point should be based on the Prospect Theory where losses and gains are discounted differently and when the reference point determines how an outcome is perceived (Kahneman and Tversky, 1979). Providing answers to these questions will require innovative interdisciplinary research methods and new forms of data (based on big-data principles) to facilitate effective policy evaluation approaches and the re-design of financial and policy instruments to address housing market instability in a repeatable framework. Existing research attempts to examine the following soft human factors:

1 risk-taking attitude (*e.g. on mortgage allowance and loan-to-value ratio*) (Bernanke, 2010);

10   Attributes of housing markets

2 secondary information processing (*e.g. processing market information and news reports*) (Wallace, 2008);
3 perceptions about present and future gains/losses (*e.g. assumptions and calculations about capital gains, job security and future income*) (Genesove and Mayer, 2001);
4 'norms' and 'social connotations' (*e.g. on homeownership vs renting in different countries*) (Després, 1991);
5 cultural factors (*e.g. perceptions of 'house' and 'home' from particular cultural lenses*) (Rykwert, 1991; Després, 1991);
6 reference points (*e.g. peer group, neighbours' and past choices*) (Clapp et al., 2008);
7 intra-family bargaining (*e.g. dominance of family members' opinions*) (Ott, 2012).

Behaviour of economic agents are difficult to predict. Even within the simple framework of demand-supply imbalances, behaviours of economic agents (i.e. consumers, investors, intermediaries) change as their market perception changes and mood swings. Buyers and sellers shape their behaviour based on 'signal processing' of the information set at a given point in time. The 'signal processing' may vary from buyers to sellers, and it depends on the phase of the market cycle they operate in. Such behaviours will be reflected by the adjustment of buyers' and sellers' reservation prices, which may vary over time. However, such behaviours can also be driven by culture, value, emotions, attitude, personality traits and neurological conditions as much as hard purchasing power (Camerer et al., 2005; Carr and Steele, 2010). Cognitive attributes and biases often prevent people from making rational decisions, despite their best efforts and self-interest. The fundamental connection between economic choices and personality traits has been severely understudied in mainstream economic science. As a result, our understanding is extremely limited, and policy prescriptions are often rendered inconsistent and frequently go awry in realising the stated objectives.

Moreover, it is reasonable to assert that most financial decisions including housing (which is the most important financial decision in most persons' lives) are *made under stress* (Porcelli and Delgado, 2009). Therefore, psychological traits such as overconfidence, projection bias and the effects of limited attention should be part of the standard economic analytics that are used to devise policies and market instruments such as taxes, subsidies, mortgage products, pension schemes, etc. Existing tools and techniques (such as multivariate regression models) in empirical economics are not fully adequate in *un-masking* and identifying the '*animal spirits*' in economic decisions, especially in the context of the housing market.

A sizable number of studies have now developed a body of knowledge on various aspects of consumer sentiment indices. Starting with Katona (1975), Mishkin (1978), Linden (1982) and others, several researchers analysed the Reuters/University of Michigan's Index of Consumer Sentiment and the Conference Board's Consumer Confidence Index (e.g. Carroll et al., 1994; Matsusaka and Sbordone, 1995).

Within real estate literature, much attention has been put on market activity data from the residential sector (Goodman, 1994; Weber and Devaney, 1996; Dua, 2008; Nanda, 2007; Croce and Haurin, 2009; Ling et al., 2015; Marcato and Nanda, 2016)[2]. On the commercial real estate sector, Baker and Saltes (2005), Clayton et al. (2009) and Marcato and Nanda (2016) analyse the role of sentiment in understanding market dynamics and,

---

2 See a more complete discussion in Marcato and Nanda (2016).

more recently, Das et al. (2014) also attempted to explain the sentiment-induced trading behaviour of institutional investors (and subsequent impact on pricing) for publicly traded real estate investment trusts (REITs) throughout different market cycles. Marcato and Nanda (2016) have analysed several real estate sentiment indices to examine current and forward-looking information content that may add useful precision to forecasting demand and supply indicators. They analyse the dynamic relationships within a Vector Auto-Regression (VAR) framework. Their findings, based on the quarterly US data over 1988–2010, largely confirm usefulness of sentiment proxies in predicting patterns in real estate returns.

Acemoglu and Scott (1994) analysed aggregate data from the UK and their findings lent support to the case of significant predictive power of sentiment data. The main question is whether consumer confidence is consistent with the Rational Expectations-Permanent Income Hypothesis (REPIH), i.e. if consumers have rational expectations, then any changes in their consumption should be unpredictable, i.e. it would follow a random walk. If the consumer confidence measures add useful information to predicting future income then it can be argued that these indices contain consumers' private information, which is otherwise not contained or revealed in any other observable economic data. However, the authors do not support such possibilities and rather put forward an argument that the confidence indicator is a leading indicator.

Bram and Ludvigson (1998) examined consumer attitudes with a focus on the forecasting ability of the Reuters/University of Michigan's Index of Consumer Sentiment and the Conference Board's Consumer Confidence Index. Their findings largely suggested varying forecasting power between these surveys. However, Souleles (2004) used underlying micro data of the Reuters/University of Michigan Index of Consumer Sentiment and found that the sentiment index could significantly raise the accuracy of the consumption growth forecast. This is useful, as the use of micro data did not lead to any aggregation bias, as in other studies. Vuchelen (2004) presented a mixed bag of evidence on whether much could be gained by using the consumer sentiment surveys. Several other studies in a cross-section of countries yielded a similar mixed bag of conclusions – Malgarini and Margani (2009) and Parigi and Schlitzer (1997) in Italy, Chua and Tsiaplias (2009) in Australia, Chang et al. (2012) in Taiwan, Easaw and Heravi (2004) in the UK, Utaka (2003) in Japan, and Fan and Wong (1998) in Hong Kong.

In the housing market, issues and related implications of information asymmetry which can lead to severe market inefficiencies have been analysed by a body of works. Case and Shiller (1989) is a leading work in this area. A number of researchers have examined the challenges of future price expectations that cannot be explained by fundamental drivers of the housing market. The focus on non-fundamental factors is important but very tricky to resolve. The challenge becomes significant in markets with infrequent transactions such as the housing market. The fact that no two houses are similar adds to the estimation biases. Nonetheless, several attempts have been made.

For example, Nanda (2007) analysed the NAHB/Wells Fargo Housing Market Index (HMI) based on a survey of home builders. He used monthly data from 1985 to 2006. The inclusion of the HMI has been shown to increase the explanatory power of the model explaining housing starts and permits. Croce and Haurin (2009) compared the Reuters/University of Michigan Index of Consumer Sentiment information on housing with the HMI and concluded that consumer sentiment measure was better in predicting housing production data than the HMI, which is interesting as HMI is based on the inputs from the homebuilding sector. With similar objectives, Ling et al. (2013) examined a number of sentiment using

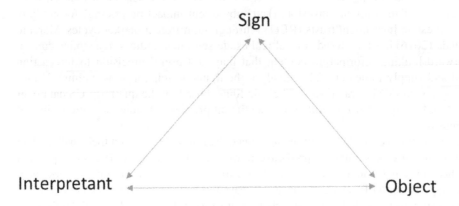

Figure 2.1 Semiotic principles: Triad of semiotics.
Source: Peirce, C.S., 1902. Logic as semiotic: The theory of signs.

surveys of home buyers, home builders, and mortgage lenders in the US. The orthogonalised (i.e. eliminating correlations with other explanatory variables) sentiment measures seemed to have better performance in forecasting house price appreciation in subsequent quarters. The results are robust to several assumptions and were derived after controlling for fundamentals and market liquidity. More recently, Jin et al. (2014) found that non-fundamental-based (irrational) consumer sentiment is a significant exogenous variable in explaining the pricing dynamics. Hohenstatt and Kaesbauer (2014) examined the Google online search query data (as an indicator of consumer interest and sentiment in the market) with a similar purpose. In a recent paper, Heinig and Nanda (2018) analysed various indicators/proxies for sentiment in the commercial real estate context.

Much of the above discussion does not also take into account the possibility that different individuals may view one attribute differently, which would influence their 'economic valuation' of the attribute. This is due to variations in sense making of housing attributes and resulting valuation or bid prices, which can be a critical element in a housing transaction.[3] It is influenced by not only the sign language of the property market, but also the 'silent language' that may influence buyers' attitude. There is not much research done on soft human factors emotion in the wider housing market to look at how buyers make sense of house attributes and prices according to the attitude, value, culture and social norms they follow. Li and Nanda (2019) apply semiotic principles to the aspects of homeownership.

Semiotics is the science of the signs that studies the whole cycle of sign from its creation, processing, to its use, with more emphasis on the effect of signs. A sign is usually used to express or communicate some aspects (e.g. picture, specification of the property features or even conversation with the agent). Therefore, a sign itself can be seen as a process in which sense-making is essential (e.g. to make sense of the property information for decision-making). This sign process or sense-making process is called semiosis (Peirce, 1958).

Figure 2.1 shows the triad of semiotics. From the housing market perspectives, property information with respect to financial and physical aspects can come under 'Sign'; actual

---

3 The following discussion is based on the working paper "Housing Market Choices: A Semiotic Analytical Framework" 2019 (Nanda, A., Li, W.).

property, value and all other aspects of the house and the market may be classified under 'Object'; however, understanding and judgement of the property's worth and attributes depends on the buyer's perception, which will determine the decision-making process and transaction price – this can fall under 'Interpretant'.

- Property description (e.g. asking price, website specification, pictures, conversation with an agent) as the sign which is considered as a signifier/vehicle/*representamen*.
- Actual property (e.g. apartment/house in the reality) as the object that is signified/described/represented.
- Interpretant: the effect of the sign on the buyer who reads and interprets the property information. This leads to a process and a result of sense making using prior knowledge, experience and culture, for example, an impression (of the property value) or an interpretation from seeing the property and by reading the sign. This impression or interpretation would depend on the buyer's personal experiences and prior knowledge. A sign may mean a different object to different persons involved in the sign process; therefore the effect or *interpretant* can be different. This subjectivity is essential in every semiosis. Therefore, the impression of the property value may differ depending on the buyer's knowledge and experience, the context, culture and convention. So the judgement of mapping between the property sign (asking price, etc.) and the actual property will never be the same.

The property value reflects and amplifies the value judgement of the owners and buyers in the market system where there are cultural systems represented by all interested parties. When looking at how a property will be judged, one needs to take into account all different judgements of all those parties that are affected. The cultural response to the property value may not be explicit but it flows from an informal process against which properties are measured and evaluated. The culture is the basic evaluative system. Its judgements are decisive but seldom fully articulated. However, failure to understand the cultural perception often results in a misplaced evaluation of house value from the buyer's perspective. Therefore, intermediaries in the property market can facilitate such processes better with a deeper understanding of an implicit evaluation system of the buyers. The properties may actually speak to a culture and the culture replies, but it speaks in direct response to actual situations and in a silent language. Table 2.1, based on the discussion in Li and Nanda (2019), provides a mapping of ten fundamental needs with housing-specific characteristics. This is based on a framework of cultural means of conveying silent messages proposed by Hall (1959).[4] These fundamental means from a housing perspective are: Subsistence, Classification, Territoriality, Temporality, Learning, Recreation, Protection, Exploitation, Association and Interaction.

These characteristics may inform the nature and structure of the drivers and factors of housing demand and supply. It follows from the earlier discussion of unique attributes that make housing a special economic good. It is important to note that there can be some overlaps among these evaluation aspects, e.g. security system features may come under protection as well as exploitation.

---

4 Hall, E.T. (1959). *The silent language.* New York: Doubleday. Also see a discussion in Stamper, R. (1988) Analysing the cultural impact of a system. *International Journal of Information Management* 8:2 107–122.

Table 2.1 Ten evaluation aspects for homeowners/householders in relation to housing choice

Aspects	General explanation	Observable parameters for evaluation
Subsistence	• Existence; survival; protection as shelter; • Future generations – overlapping generations in terms of consumption and investment needs	• Consumption motive – number of bedrooms, bathrooms (as minimum shelter requirement) • - Investment motive – expected capital growth beyond inflation
Classification	• A house may reflect one's identity, status – e.g. viewed as a conspicuous good	• Buyer's personal attributes – age, education, marital status, sex, ethnicity
Territoriality	• Spatial fixity of a house • Privacy of the occupant • Sense of ownership of the land, objects, associated pride • Exclusivity of human existence	• Floorplan • External land usage – garden, garage, fence, boundary walls • Location choice based on physical/geographic/spatial parameters
Temporality	• Age of the house as a durable good – patterns of demand evolving over the age of the house • Age of the occupant	• Intended tenure duration • Time of re-sale
Learning	• Enhancement of individual's life experience through housing choices • Technical learning (process defined – tools, legal, finances; social learning)	• 'Housing Career' along with Life Cycle stages – moving from parental housing->*student housing->renter->first home-owner->buyer as well as seller->second home-owner->retirement home*
Recreation	• Fulfilment of life's non-survival aspirations facilitated by homeownership or occupation, e.g. use of objects, upkeep of surroundings	• Interior decoration • Property decoration • Added aesthetics • Use of objects
Protection	• Fundamental provider and sense of security from elements, people, danger within the confinement of home – as a 'shelter'	• Security system • Locks/doors/windows/fence • Crime rate consideration
Exploitation	• Use of the home space through physical objects and virtual linkages to satisfy needs of human existence • Different patterns in housing consumption and interaction based on occupant being on different part of the life cycle	• Television/dish antenna • Internet/Wi-Fi • Double-glazed window/solar panels • Advanced alarm system/ CCTV • House usage/occupancy level – over-consumption/under-consumption
Association	• A house helps individual in associating oneself to the surrounding – the nature, the neighbour, the network (social, virtual)	• Choice of community/neighbourhood • Choice of location based on demographic factors
Interaction	• Using the spatially fixed house to interact within, through and external linkages of the house – human-to-human, human-to-objects interactions during the transaction process and consumption of the house	• With agent, bank • With friends, colleagues • With family (intra-family bargaining)

Source: Li and Nanda (2019)

In this chapter, I have discussed the basic aspects and attributes of the housing market with reference to unique characteristics, implications of consumer sentiment and culture through multi-disciplinary perspectives. In the next chapter, I delve into the economic fundamentals of the housing market.

Selected research topics

- Understanding the factors that contribute to spatial and temporal variations in durability in housing stock.
- Determining the factors that make the housing market inefficient.
- Impact of technology on housing market inefficiencies.
- Understanding links of cultural issues with housing choices.

3 Economy, housing demand and supply

Chapter outline

- Simple model of the economy
- Housing in the economy
- Demand and supply of housing
- Sources of bias in estimation
- New forms of data
- Selected research topics

Economic analysis is about understanding forces and activities that shape markets and sources of changes in willingness of households and firms to engage in economic activities. A key challenge is to decipher how these lead to a set of choices that households and firms can select. While there are many other players in the economy, there are two main protagonists – households who provide labour and capital, and firms who use those labour and capital to produce goods and services that households purchase. This is a very simple circular model of the economy. Much of this idea is based on neoclassical thoughts which got refined by institutional economics bringing in roles of rules and institutions. The model is refined with principles of information economics where several aspects of information asymmetry have been examined variously and incorporated by the researchers.

Households play a key role in the circular flow of the economy model (see Figure 3.1) as the supplier of inputs (labour and capital) and demanders of the outputs. Any friction happening at any part of this chain will result in economic fluctuation in the form of expansionary (e.g. boom, increasing aggregate demand) or recessionary (e.g. falling aggregate demand) pressures. The most common measure of size of the economy is Gross Domestic Product (GDP). **Gross Domestic Product (GDP) is the value of all *final* goods and services produced *within a country* in a *given period* of time.**

$$GDP = Y = C + I + G + NX \qquad (3.1)$$

Consumption (C), Investment (I), Government Purchases (G), Net Exports (NX), i.e. (Exports – Imports).

Housing appears in different ways in the GDP formula. It may be part of the consumption or investment depending on new or existing housing goods and services.

- **Consumption (C):** The spending by households on goods and services, except purchases of new housing.

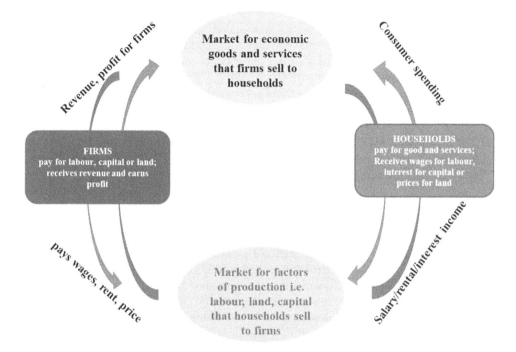

Figure 3.1 A simple model of the economy: Circular flow diagram.

Source: Adapted from Essentials of Economics 3e by N. Gregory Mankiw, South-Western.

- **Investment (I):** The spending on capital equipment, inventories and structures, including new housing.
- **Government Purchases (G):** The spending on goods and services by local, state and federal governments. This does not include transfer payments because they are not made in exchange for currently produced goods or services.
- **Net Exports (NX):** Exports minus imports.

GDP estimates the monetary value of final goods and services. Although GDP includes mostly market goods and services, it may also include several non-market goods and services, e.g. defense expenses or public or government-funded education services. GNP or Gross National Product is an alternative concept which takes into account outputs of the residents of a country. Foreign output by a domestic organisation is not included in GNP. GDP measure is limited in several ways:

- It does not take into account informal productions such as voluntary work, household work, family help or the underground economy.
- It does not take into account depreciation of machinery or buildings or any capital stock. If the depreciation is deducted, it is called the Net Domestic Product.
- It does not measure wellbeing or happiness or standard of living or quality of life. Although GDP growth may indicate more productive use of resources, employment and income, it may come with costs (such as environmental or human development costs). UN's Human Development Index attempts to capture those.

18 Economy, housing demand and supply

The GDP can be defined in three different ways:

1 The **production approach** sums the 'value-added' at each stage of production, where value-added is defined as total sales less the value of intermediate inputs used during the production. For example, a house is the final product and an architect's services would be an intermediate input.
2 The **income approach** sums the incomes generated by production. For example, the salary or compensation employees receive for building the house and the operating surplus of house building companies.
3 The **expenditure approach** adds up the value of purchases made by final users such as households, firms and government organisations. In terms of the housing example, there can be many such expenditures.

For example, Figure 3.2 shows the UK GDP over the years. It shows the economic cycles for the UK economy.

GDP is one of the most consistently available economic data available across all economies of the world. It is usually estimated and published by the national statistical agencies. There are international standards that are followed in estimating GDP. The international standard for measuring GDP is contained in the System of National Accounts, 1993, compiled by the International Monetary Fund, the European Commission, the Organization for Economic Cooperation and Development, the United Nations and the World Bank.

GDP is the key measure for examining fluctuations in economic performance. The latest available GDP measure (latest quarter) is often revised in the next period as the complete data is accounted for. Figure 3.3 shows GDP figures for a selection of countries and regions in the world. High income and Organization for Economic Cooperation and Development (OECD) countries show the largest share in the total GDP of the world.

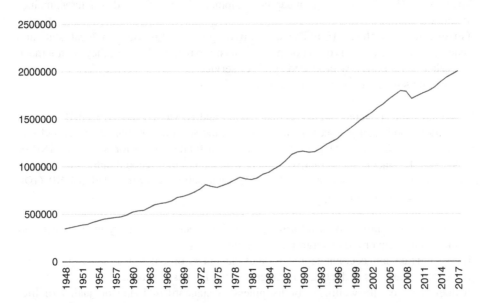

Figure 3.2 United Kingdom GDP (Gross domestic product: Chained volume measures: Seasonally adjusted £m).

Source: ONS.

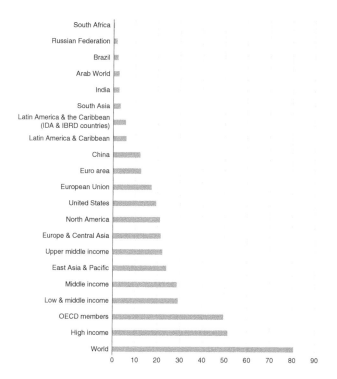

Figure 3.3 GDP (current US$ trillions, 2017) across the countries and regions.

Source: Based on data from World Bank; for a full list, refer to the *source*.

BRICS (Brazil, Russia, India, China and South Africa) countries are also shown in Figure 3.4 with their 2017 GDP figures.[1]

Two of the largest countries in terms of population, India and China, show significant growth in their GDP figures, especially in the last three decades (see Figures 3.5 and 3.6).

Globally, there is a huge variation in GDP performance across countries. The following chart from IMF shows the annual per cent change in 2018. While there are pockets of strong growth, there are several regions which are not growing, or expanding at a much slower rate than other leading countries like India and China. Much of the GDP map is changing fast due to variation in regional performances. Moreover, sectoral patterns in terms of contribution to GDP have also been changing and are expected to change. Internet and digital technology are influencing those patterns.

The National Bureau of Economic Research (NBER) provides the official dates for recession in the US since 1857. The chronology provides the dates of peaks and troughs in economic activity. According to the NBER website (https://www.nber.org/cycles/recessions.html),

> a recession is a period between a peak and a trough, and an expansion is a period between a trough and a peak. During a recession, a significant decline in economic

---

1 BRICS is the acronym given by Jim O'Neill for five major emerging national economies: Brazil, Russia, India, China and South Africa.

activity spreads across the economy and can last from a few months to more than a year. Similarly, during an expansion, economic activity rises substantially, spreads across the economy, and usually lasts for several years. In both recessions and expansions, brief reversals in economic activity may occur – a recession may include a short period of expansion followed by further decline; an expansion may include a short period of contraction followed by further growth.

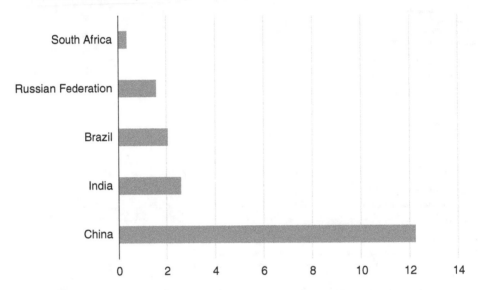

Figure 3.4 GDP (current US$ trillion, 2017) – BRICS.

Source: Based on data from World Bank.

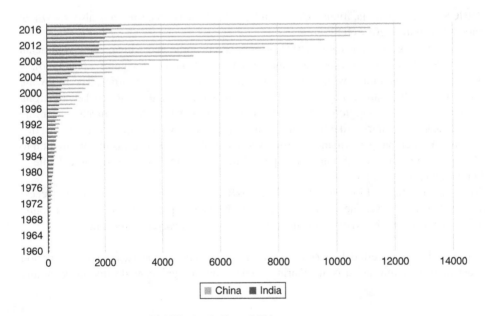

Figure 3.5 GDP (current US$ billion) – India and China.

Source: Based on data from World Bank.

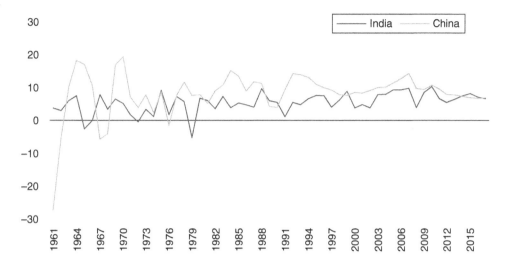

Figure 3.6 GDP growth (annual %) – India and China.
Source: World Bank.

Based on these definitions, the Committee applies its judgement in order to determine on a case-by-case basis whether a contraction or expansion is only a short interruption to the continuing trend or a major disruption. This is important and shows that cycles may not have much commonalities which needs to be taken into account. Otherwise, the judgement can go wrong. For example, as noted in the website, in 1980–1982, the Committee determined that the contraction that began in 1981 was not really a continuation of the one that began in 1980, but categorised it as a separate full recession. Moreover, the Committee does not have a fixed definition of economic activity. It examines and compares the behaviour of various measures of broad activity: real GDP measured on the product and income sides, economy-wide employment and real income. The Committee may also take into account the indicators that do not cover the entire economy, such as real sales and the Federal Reserve's index of industrial production (IP). It further notes that

> the Committee's use of these indicators in conjunction with the broad measures recognizes the issue of double-counting of sectors included in both those indicators and the broad measures. Still, a well-defined peak or trough in real sales or IP might help to determine the overall peak or trough dates, particularly if the economy-wide indicators are in conflict or do not have well-defined peaks or troughs.

Table 3.1 shows the dates.

Contribution of housing to GDP

Housing is a key sector of any economy. It's inter-linkages with the rest of the economy are wide and deep through various other sectors. It creates very significant amounts of employment and income streams for various economic agents. If the housing sector is excluded

22 Economy, housing demand and supply

Table 3.1 US recession dates: Business cycle reference dates since World War II

Peak	Trough	Contraction	Expansion
Quarterly dates are in parentheses		Peak to trough	Previous trough to this peak
February 1945 (I)	October 1945 (IV)	8	80
November 1948 (IV)	October 1949 (IV)	11	37
July 1953 (II)	May 1954 (II)	10	45
August 1957 (III)	April 1958 (II)	8	39
April 1960 (II)	February 1961 (I)	10	24
December 1969 (IV)	November 1970 (IV)	11	106
November 1973 (IV)	March 1975 (I)	16	36
January 1980 (I)	July 1980 (III)	6	58
July 1981 (III)	November 1982 (IV)	16	12
July 1990 (III)	March 1991 (I)	8	92
March 2001 (I)	November 2001 (IV)	8	120
December 2007 (IV)	June 2009 (II)	18	73

Source: Based on information from NBER; for a full analysis and more information, refer to the source. http://www.nber.org/cycles.html

from the GDP calculation, it can reduce GDP growth by a significant amount. Especially with widespread linkage with the financial sector, the housing sector plays a crucial role in fuelling the economic engine.

Let's look at the housing's contribution to two large economies – US and UK. Figure 3.7 shows a long-run view of housing's share of US GDP. From 1980 to 2007/8, the share was hovering around 17–18%. Due to the Global Financial Crisis (GFC) and severe housing sector turmoil, the share dropped to 15–16% post-2008. In the world's largest economy, these

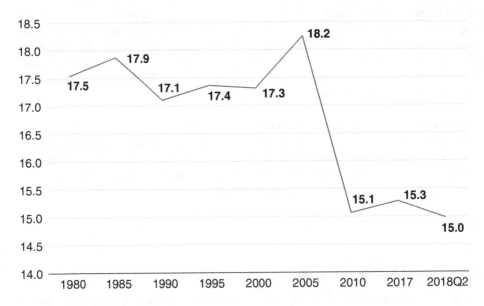

Figure 3.7 Housing's contribution to US GDP.

Source: Based on data from NAHB.

shares are very significant. Any weakness in this sector can erase a significant portion of the economic performance.

Table 3.2 and Figure 3.7 show the contribution of residential investment to GDP before the recession in 2008 and afterwards as computed by the National Association of Home Builders in the US. Several developed countries have been studied and a very significant contribution of the housing sector has been reported with US leading the way where construction sector is one of the biggest employers. The bottom panel of the table shows the potential impact on the GDP performance if the housing sector is taken out. It is important to note that such calculations are not easy and care must be taken to avoid double-counting.

The National Association of Home Builders in the US also estimates the economic impact that residential construction has on the US economy. The national estimates for 2014 include the following:

- Building an average single-family home: 2.97 jobs, $110,957 in taxes.
- Building an average rental apartment: 1.13 jobs, $42,383 in taxes.
- $100,000 spent on remodelling: 0.89 jobs, $29,779 in taxes.

The jobs are given in full-time equivalents (full-time equivalent is enough work to keep one worker employed for a full year based on the average hours worked per week in the relevant industry. Table 3.3 shows the detailed figures for single-family houses, multi-family houses and remodelling.

Table 3.2 Housing's contribution to GDP performance (US)

	Housing's contribution to GDP				
	1985	1995	2005	2010	2018Q2
Constant dollars (2009, millions)					
Gross domestic product	7,951,074	10,630,321	14,912,509	15,598,754	18,507,200
Residential fixed investment	423,294	502,646	885,391	383,023	613,662
Housing services	997,454	1,343,244	1,835,341	1,966,848	2,158,539
Residential fixed investment + housing services	*1,420,748*	*1,845,890*	*2,720,732*	*2,349,870*	*2,772,201*
Percentage of GDP					
Residential fixed investment	5.3	4.7	5.9	2.5	3.3
Housing services	12.5	12.6	12.3	12.6	11.7
Residential fixed investment + housing services	*17.9*	*17.4*	*18.2*	*15.1*	*15.0*
Contribution to real GDP growth					
Gross domestic product growth	4.2%	2.7%	3.3%	2.5%	4.1%
Housing services	*0.45%*	*0.36%*	*0.49%*	*0.16%*	*0.30%*
Residential fixed investment	*0.11%*	*–0.15%*	*0.41%*	*–0.07%*	*–0.04%*
Government expenditures and investment	–0.1%	–0.1%	–0.1%	–0.1%	0.4%
Net exports of goods and services	–1.0%	–0.4%	0.0%	1.5%	1.1%

Source: Adapted from NAHB; for a full analysis, refer to the source. https://www.nahb.org/en/research/housing-economics/housings-economic-impact/housings-contribution-to-gross-domestic-product-gdp.aspx

Table 3.3 Income/employment impacts of residential construction on the US economy

	Full time equivalent jobs	Wages and salaries	Wages and profits combined
Per new single-family home			
All industries	2.97	$162,080	$280,433
Per new multifamily rental units			
All industries	1.13	$60,877	$107,715
Per $100,000 spent on remodelling			
All industries	0.89	$48,212	$83,402

Source: Based on data from NAHB; for a full list, refer to the source. https://www.nahb.org/en/research/housing-economics/housings-economic-impact/impact-of-home-building-and-remodeling-on-the-u-s--economy.aspx

NAHB cautions that these national estimates are designed for use when the impacts on all US suppliers of goods and services to the construction industry – for example, manufacturers of building products – are of particular interest. The national estimates should not be used to try to analyse economic impacts confined to the state or local area where the housing is built. NAHB has a separate Local Economic Impact section on its web site for that.

If we take a look at the UK housing market, according to the estimates by the Office of National Statistics (ONS), the value of the household sector increased by £750 billion in 2016, making it the greatest contributor to the change in UK's net worth. Insurance, pension and standardised guarantee schemes rose by £348 billion, out of which almost £319 billion accounts for the pension schemes. A large driver in household growth is the value of underlying land, which has appreciated by £266 billion in 2016. In comparison, the value of dwellings excluding land has gone up by £51 billion. From 1995, the rate of growth of land value has far outstripped the growth of dwellings value (Figure 3.8).

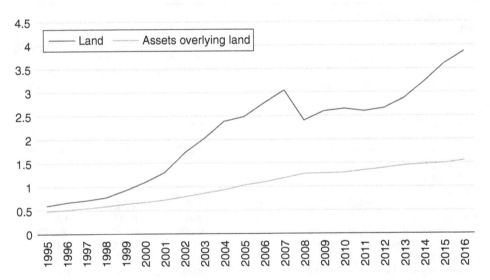

Figure 3.8 Household's sector value of land and its overlying assets, 1995 to 2016 (£ trillion).

Source: Based on data from ONS, UK. https://www.ons.gov.uk/economy/nationalaccounts/uksectoraccounts/bulletins/nationalbalancesheet/2017estimates

The multiplier effect

The contribution to GDP can be understood by the multiplier effect (refer to standard economics textbooks for detailed exposition for the following derivation).
From equation (3.1), for an open economy, GDP = Y = C + I + G + NX
For a closed economy with no foreign trade, i.e. nil NX, GDP = Y = C + I + G
We know, Investment (I) = Savings (S); and G = Tax (T)
i.e. Y = C + S + T; and Disposable Income = Y_d = Y – T = C + S
So, if C = C_0 + cY_d
Then, S = -C_0 + (1-c)Y_d = S_0 + sY_d
where c = marginal propensity to consume; s = (1-c) = marginal propensity to save
Note that: mpc + mps = 1
Y = C + I + G; but C = C_0 + c(Y-T)
So, Y = C_0 + c(Y-T) + I + G
i.e. Y = C_0 + cY – cT
i.e. Y – cY = C_0 + I + G – cT

Y – cY = (1-c)Y = C_0 + I + G – cT

$$\text{i.e.} \quad Y^* = \frac{1}{1-c}\left[C_0 + I + G - cT\right] \quad (3.2)$$

- 1/(1-c) is called the aggregate expenditure multiplier or autonomous expenditure multiplier.
- It is positive.
- An increase in autonomous spending has an amplified impact on GDP.

Suppose, mpc_{uk} = 0.6; what would be the impact on GDP if Govt. spends £1bn in infrastructure investment?

$$\Delta Y = \frac{1}{1-c}\Delta G \quad (3.3)$$

Marginal propensity to consume (MPC) and marginal propensity to save (MPS) are very relevant and significant concepts for the housing market. These determine our ability to pay for deposit constraints and demand for mortgages and property types.

In the following sub-section, some key economic concepts are explained that are crucial for a rigorous economic analysis of housing market dynamics.

Demand for housing

One of the main economic goods that households require is housing. Two key concepts in economics regarding provision of goods and services are 'demand' and 'supply'. Demand is about willingness to pay and ability to purchase at certain prices. It clearly links to an individual's circumstances such as preferences, financial strength and timeframe. **Demand or quantity demanded is the amount of an economic product or service that a consumer would want to buy at a given price. The demand curve is usually downward sloping, since consumers may want to buy more as price decreases (law of demand) (Figure 3.9).**

In case of the real estate market, a consumer may demand (or a producer may supply):

- **Land** (transaction is about the rights to use.)
- **Physical property** – house, office, warehouse or any built space.
- **Financial instruments** like stocks of a real estate company, mortgage-backed securities.
- **Services** – property management services, advisory services.

Demand in economics is a key concept. In the context of the housing market, the demand can be defined in terms of physical home and related products and also services and financial products such as mortgages and home insurance plans.

Violating the law of demand

Veblen goods are a group of commodities for which demand increases as the price increases, as greater price implies greater status, instead of decreasing according to the law of demand. A Veblen good is also a positional good (e.g. expensive special edition cars, very expensive homes).

A **Giffen good** is one for which demand and consumption as price rises, violating the law of demand. The classic example given is of inferior quality staple foods. As the price of the cheap staple rises, due to a lack of substitutes, the income effect dominates and consumers tend to buy more of the good, leading to rising prices.

In case of the housing market, future price expectations play very important roles in shaping up the current period's demand, which can lead to violation of standard demand relationship and anomalies.

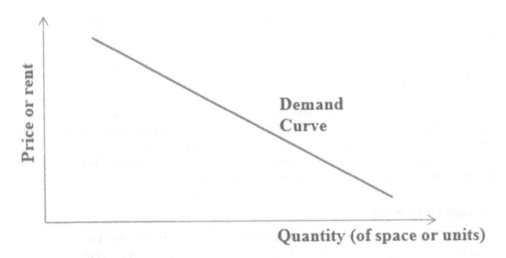

Figure 3.9 Demand curve.

Economy, housing demand and supply 27

Supply or quantity supplied is the amount of an economic product or service that a producer would want to sell at a given price. The supply curve is usually upward sloping, since producers may want to sell more as price increases (Figure 3.10).

But in the real estate space market, supply is fixed in the short-run. So, the supply curve is vertical (see Figure 3.11). This is due to the fact that it requires a significant amount of time to gather all resources, go through all processes and seek planning permissions. This time can be significant and does vary over time and space.

Let's take a look at the stock of housing across a number of countries. Table 3.4 reports the total and per 1000 population number of dwellings in 2010 and 2015 across several countries.

Market equilibrium refers to a condition where a market price is determined through competition such that the quantity demanded is equal to the quantity supplied. In Figure 3.12, the market equilibrium is obtained at the intersection (P*, Q*) of demand and supply curves.

But, in the real estate space market, supply can be fixed in the short-run. This is a very important aspect in determining the extent of price variation. So, in the short-run, price is determined by the shift of the demand curve if there is any demand shock (+ve or –ve). Due to the vertical short-run supply curve, the price change (P* to P~) is equal to the extent of the shift of the demand curve (Figure 3.13). This creates short-run fluctuations and can occur often in the economy. Since instantaneous supply adjustment is not possible, the price effect can linger for a while until supply increases, keeping everything else constant.

There is a clear distinction between movement along a demand/supply curve, and a shift in a demand/supply curve. A change in price causes the quantity demanded/supplied to change, i.e. movement along the curves. When a non-price determinant of demand/supply changes, the curve shifts.

Some demand shifters are:

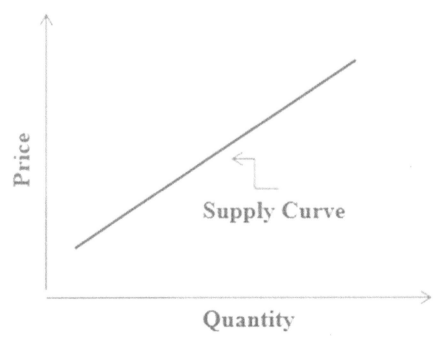

Figure 3.10 Supply curve.

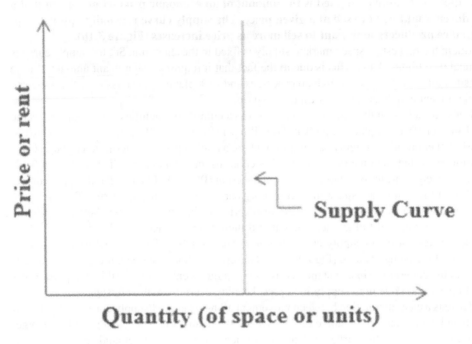

Figure 3.11 Short-run supply curve.

- changes in disposable income;
- changes in tastes and preferences;
- changes in expectations;
- changes in the prices of related goods (substitutes and complements);
- population size and composition.

A key tool to understand demand and supply fluctuation is the concept of elasticities. Supply is relatively more inelastic due to the fact that it takes a significant amount of time to build houses. Price elasticity of demand is a measure of the sensitivity of the quantity to changes in the price. Elasticity measures how much the quantity will change in percentage terms for a 1% change.

- If the price elasticity is between zero and 1, demand is said to be inelastic.
- If the price elasticity equals to 1, the demand is unitary elastic.
- If the price elasticity is greater than 1, demand is elastic.
- A high elasticity indicates that consumers will respond to a price rise by opting for less amount of the good and that consumers will respond to a price cut by buying more.

Income Elasticity of Demand measures the responsiveness of the quantity demanded for a good or service to a change in the **income levels**. It is calculated as the ratio of the percentage change in quantity demanded to the percentage change in **income**. Both price and income elasticity of demand are very important concepts for the housing market. These change over time and across regions and countries. Increases in **income** can lead to a high percentage increase in **demand**. As their **income** levels improve, many individuals may switch from one tenure to another tenure or upgrade within the same tenure or even seek additional properties.

At the fundamental basis of demand theory, there is the concept of utility (refer to standard economics text books for detailed exposition for the following derivation). Individual utility is maximised when the ratio of the marginal utilities (μ_H, μ_C) equals the ratio of the prices of H & C (P_H, P_C)

$$\mu_H / \mu_C = P_H / P_C \tag{3.4}$$

The individual household demand for housing is a function of income and relative price of housing

$$H_i^d = f(Y_i, P_H / P_C) \quad i = 1, \ldots, N \tag{3.5}$$

Total demand for a given number of households (N) is the sum of the individual demand cases.

$$H^d = N * f(Y_i, P_H / P_C) \tag{3.6}$$

Figure 3.14 shows the indifference curve and utility maximisation point identified by equation (3.4).

Table 3.4 Total housing stock in OECD and EU countries, selected years

	2010	2015	2010	2015
Australia	9,116,942	9,615,800	411	401
Austria	4,477,000	4,506,000	530	525
Canada	14,618,960	..	428	..
Chile	4,808,645	5,079,233	280	290
Denmark	2,597,968	2,628,338	463	464
France	33,484,000	34,923,000	533	546
Germany	40,630,300	41,446,300	506	510
Greece	6,371,901	..	573	..
Hungary	4,408,050	4,420,296	441	449
Ireland	2,019,000	2,022,000	440	437
Italy	31,208,161	..	526	..
Japan	60,628,600	..	477	..
Korea	17,738,800	19,559,100	357	383
Mexico	35,617,724	..	314	..
Netherlands	7,290,000	7,588,000	440	449
New Zealand	1,500,000	1,803,400	339	398
Norway	2,205,191	2,316,647	437	448
Poland	13,496,000	13,983,000	350	363
Portugal	5,859,540	5,926,286	554	571
Spain	25,208,623	..	540	..
Sweden	4,633,678	4,637,636	485	476
Switzerland	4,234,906	4,289,428	527	527
United Kingdom	27,448,000	28,073,000	439	436
United States	131,704,730	134,789,944	416	419

(Number of dwellings, total and per thousand inhabitants, 2010 and 2015)
Source: Based on data from OECD Questionnaire on Affordable and Social Housing; for more information, refer to the source.

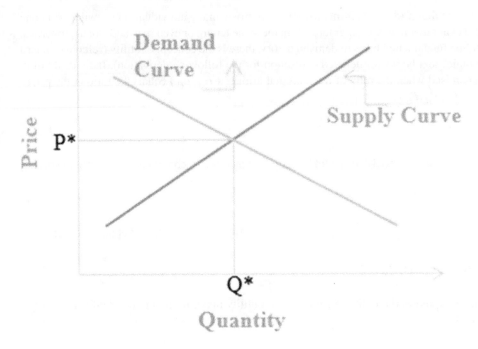

Figure 3.12 Market equilibrium.

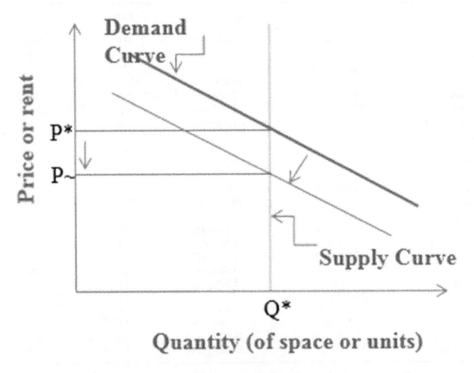

Figure 3.13 Market clearing adjustments.

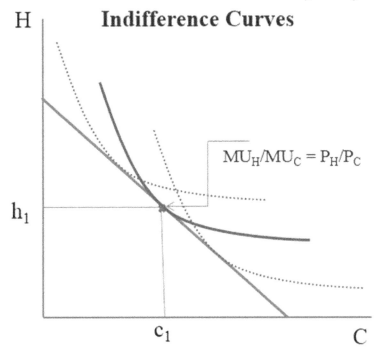

Figure 3.14 Indifference curves.

Assume we only consume two goods: housing (H) and another non-housing composite good (C). Suppose our monthly house rent is £500. The local government has just introduced a rent subsidy for students that reduces the cost of housing by 20% from January 1, 2019. This implies that we have £500 × 0.20 = £100 more to spend each month. This £100 can be viewed in two different ways:

(a) Referring to Figure 3.15, budget constraint rotates to AB', increasing the housing consumption: price of housing relative to all other goods has been reduced. This can lead to more housing and less of other goods – *Substitution Effect*.
(b) The rent subsidy of £100 has actually increased our disposable income per month, i.e. more money available for buying housing and other goods, which is called the *Income Effect*.

Next we try to characterise the aggregate housing demand function. In general, literature suggests (see Meen, 2001; Miles, 1994 and others) the following core short-run drivers of aggregate housing demand:

- Income (I)
- Price of housing relative to other goods $(p_H)/(p_C)$
- Interest rates (r)
- Credit availability (Cr)
- Wealth (W)
- Tax structure of housing (t_H)

32 Economy, housing demand and supply

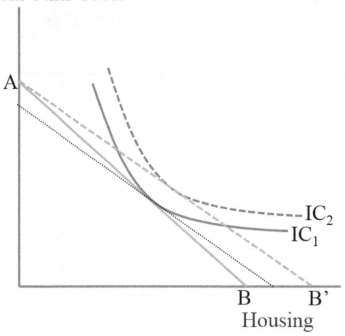

Figure 3.15 Price determination.

- Expected capital gains ($p_H e$)
- Household size (*hs*)
- Employment (*emp*)

The long-run drivers of housing demand are:

- Population size (N)
- Demographic structure and composition (D)
- Headship rates (h)
- Migration (m)

If we bring all the demand shifters into a single framework, we get the aggregate housing demand:

$$H^d = f(I, p_H, r, p_C, Cr, W, tH, pH^e, hs, emp, N, D, h, m) \tag{3.7}$$

If we have time series and/or cross-sectional data on all these components, we can effectively estimate a demand function for any city or country or region. Moreover, we can answer questions regarding the contributions made by various factors, which would differ markedly over time and across locations.

> Exercise: what are the expected signs of the factors when we estimate the above equation empirically, e.g. in a regression model?

Next we characterise the aggregate supply function. Aggregate supply function may comprise of the existing stock and new housing. Typically, new build is a small proportion of total supply.

$$H^s = (1-\pi)H_E^s + H_N^s \qquad (3.8)$$

where: H_E^s is the existing housing stock
π is the depreciation rate
H_N^s is new housing supply

$$H_N^s = G(cc, lc, r) \qquad (3.9)$$

Determinants of new supply:

cc is building material costs
lc is the cost of developable land
r is interest rate

Since we have characterised the basic structure of the demand and supply sides of the market, we can now bring those two together to examine market equilibrium. At equilibrium, aggregate demand would be equal to the aggregate supply. **Market Equilibrium Equation is defined as a set of market clearing prices and quantities, i.e. equating demand and supply side:**

$$H^d = H^s \qquad (3.10)$$

$$f(I, p_H, r, p_C, Cr, W, tH, p_H^e, hs, emp, N, D, h, m) = (1-\pi)H_E^s + G(cc, lc, r) \qquad (3.11)$$

Solving for equilibrium prices:

$$(p_H / p_C)^* = f(I, r, Cr, W, tH, p_H^e, hs, emp, N, D, h, m, H_E^s, H_N^s) \qquad (3.12)$$

$$(p_H / p_C)^* = f(I, r, Cr, W, tH, p_H^e, hs, emp, N, D, h, m, H_E^s, \pi, cc, lc) \qquad (3.13)$$

One of the key aspects of the housing demand and supply is the desirability of a location. The desirability of a location can depend on many factors. One of the key factors is transport links as mobility and ease of commuting are key considerations. The above equations provide a basic price model that can be estimated with appropriate data inputs. While this can be estimated for any market at any time, the availability of various inputs in terms of length, breadth and depth of the information would vary significantly. Many parts of the world, especially the developing regions, do not have consistent data series on most of the standard house price model inputs. Although much effort are being put on collecting and compiling data, the short history and inconsistencies (such as measurement errors) make it challenging to estimate price models.

Implications of 'equilibrium' in housing market

We often analyse the movement towards a set of market clearing house prices and quantities. It is important to analyse the movement as the market is going through a continuous adjustment process in an attempt to match demand and supply. However, the factors or determinants affecting the equilibrium point are also evolving and changing over time and space (e.g. cities and towns). However, the speed and extent of adjustment differ as well. Some markets at certain times adjust faster than other times; some do not. This depends on the constraints such as regulatory constraints and other time and space-specific factors that are not easily observed in the data. Lags in this adjustment process with slow processes and delayed impacts imply that there are short-run and long-run equilibrium. A set of pertinent questions are: How short is the short-run? How long is the long-run? It really depends on the specific context and location.

A short-run adjustment process means that prices can easily change for a number of periods in the upswings and downturns. However, as always, the market forces attempt to correct for the short-run imbalances or fluctuations or demand-supply gaps, and try to move towards a long-run equilibrium, which is a completely dynamic process with many short-run phases and events. The nature and pattern of these short-run disequilibria also vary over time and space. Since housing is the most common wealth creating vehicle for households, the future price expectation matter significantly in shaping the current period's demand and thus, the extent of short-run disequilibrium.

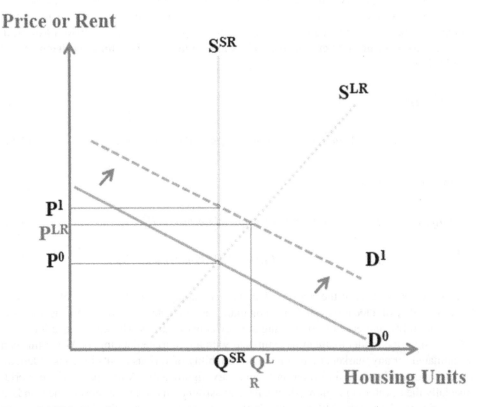

Figure 3.16 Market adjustments.

Figure 3.16 shows supply adjustments in the short-run and long-run. Supply takes time to increase or decrease. Price increase can persist longer, especially if the long-run supply is also relatively inelastic. If the long-run supply is elastic, prices will adjust accordingly and may not rise in the long-run. In Figure 3.16, the initial demand curve (D^0) shifts upwards to D^1 due to a change in some demand shifters. With a vertical short-run supply curve (S^{SR}), the price goes up to P^1. As the long-run supply curve is more elastic (S^{LR}), the price level adjusts to P^{LR}. The extent of this adjustment depends on the elasticity of the supply curve.

With all the discussions so far, a key question is: why do we study the housing market? The fundamental drivers of housing demand can also be analysed in a very long-run context. Robert Shiller, in his book *Irrational Exuberance* in 2005, analysed housing market dynamics over more than a century – 1890–2005 – for the US. In Figure 2.1 in the book, he showed a very long-run view of the US economy with fundamental factors of housing demand – population, cost of funding and construction cost. While the parity among all these factors are more or less maintained over a long time, the very marked deviation from that long-run trend is observed shortly before the 2008 Global Financial Crisis, when the house price deviated very sharply from those economic fundamentals. Housing has become a more and more important aspect of overall economic performance. With close links with the labour market, manufacturing sector and financial sector, weaknesses or strengths of the housing market can determine the economic trajectories.

In summary, housing can affect several aspects and can also be influenced by several factors in the economy.

- **Housing can significantly determine economic growth.** Housing is a complex product requiring various skillsets and several industries. Moreover, due to income effect, it can have an impact on aggregate demand both from the production of housing as well as the consumer sector.
- **Housing and labour market decisions are often linked with much implication for labour market dynamics and wages.** Labour force availability gets influenced by varying levels of cost of living across places.
- **Housing can be a significant factor for shaping migration patterns.** International, inter-regional and intra-regional (move to suburbs) movement of people are often determined by housing market issues.
- **Housing has a close link with demographics.** This can have impacts on community formation. Moreover, housing needs being associated with key life-cycle events, it can affect household formation and family structure.
- **Housing finance system is a key component of the broad financial sector.** Credit availability, cost of financing (i.e. interest rates) and level of leverage have significant bearing on the market stability.

Comparison of housing with other real estate sectors

Housing is quite different from other real estate sectors such as office, retail, industrial and hospitality (hotels and tourism). The drivers of housing demand differs in several ways from those of other sectors. Table 3.5 lists the most prominent drivers of different real estate sectors along with those of the housing market. There is no doubt that some commonalities across these sectors (e.g. interest rate or cost of funding) would impact all real estate sectors.

Table 3.5 Main demand drivers in major real estate sectors

Property type	Demand drivers
Residential single-family (owner-occupied)	• Population • Household formation • Interest rate • Employment growth (business and professional)
Office	Employment in office occupations: • Finance, Insurance, Real Estate (FIRE) • Business and professional services • Legal services
Retail/shopping	Aggregate disposable income Aggregate household wealth Traffic volume (specific sites)
Hotel, tourism-hospitality	• Air passenger volume • Tourism receipt or number of visitors
Industrial/warehouse	• World demand for consumer goods • Manufacturing employment • Transportation employment • Airfreight volume • Rail and truck freight volume

Source: Based on information from Geltner, D., Miller, N.G., Clayton, J. and Eichholtz, P., 2001. Commercial real estate analysis and investments (Vol. 1, p. 642). Cincinnati, OH: South-western. See exhibit 6–4.

Housing market area

A key challenge in understanding housing market dynamics is the identification of relevant geographic area. Although several public amenities are governed by administrative boundaries, given the dynamic changes in the housing demand shifters, those are often outdated for strategic planning purposes.

Sub-regional housing market areas are geographical areas defined by household demand and preferences for housing. They reflect the key functional linkages between places where people live and work. Research has established that changes in the house price of a region may transmit to its neighbouring and surrounding regions. The transmission mechanism may follow spatial and temporal diffusion processes. Nanda and Yeh (2014) investigated such regional housing market dynamics and interactions among local housing sub-markets in Taipei using a panel data framework and spatial panel models using annual data on median residential land prices from 41 Taipei sub-markets over the period from 1992 to 2010.[2] The empirical analysis revealed a strong effect that the spatial dependence plays a significant role in interactions among regional housing markets. The results in the study, when tested across alternative specifications, are strongly robust even after addressing unobserved time and spatial effect. Similarly, Nanda and Yeh (2014) studied spatial linkages among the US metropolitan housing markets.[3] The results confirmed strong spatial dependence. These findings provide valuable insights and have significant implications for

---

2 "Spatio-Temporal Diffusion of Residential Land Prices across Taipei Regions" 2014 (Nanda, A., Yeh, J.) *SpringerPlus* 3:505. doi:10.1186/2193-1801-3-505. http://www.ncbi.nlm.nih.gov/pmc/articles/PMC4174549/
3 "Spatio-Temporal Diffusion of US Metropolitan House Prices" 2014 (Nanda, A., Yeh, J.) Conference proceedings, Regional Studies Association, Winter conference, 2014. ISBN: 978-1-897721-49-0; 140–149.

urban strategic spatial planning and efficient use of public resources, especially mega-urban or city-region areas.

There are several advantages to regions and local authorities working together to identify sub-regional housing market areas:

1. The ability to better understand how housing markets work.
2. The ability to develop a more strategic approach to housing.
3. It can facilitate better integrated planning and housing policy.
4. The ability to pool resources and developing best practice.

How do we identify sub-regional housing market areas?

Consumers' economic interests often overlap administrative borders (e.g. shopping mall, schools, employment centres, etc.). Therefore, amenities, investments, activities and developments in one area will affect consumers' decisions in other surrounding areas. This can result in significant spillover and leakage effects. Any sub-regional policies (e.g. infrastructure development, transport network) need to recognise such effects and be implemented across the relevant regions for more cohesive effects. Moreover, it is also a very dynamic issue as interactions may change over time due to population growth and place-based transformations. Therefore, it is important to identify the area that is more representative for the policy actions. There are several major sources of information that can be used to identify sub-regional markets:

- *House price changes*, which may reflect household demand and preferences for different sizes and types of housing in different locations.
- *Household migration and search patterns*, reflecting preferences and the trade-offs made when choosing housing locations with different characteristics, e.g. identifying areas with a relatively high proportion of intra-regional moves.
- *Contextual data, such as travel to work areas*, which reflect the functional relationships between places where people work and live with economic interests.

Housing data sources

All the above attributes, demand and supply drivers and identification of housing market areas require consistently collected data items to be able to identify their contributions and roles in the housing market dynamics. Countries and regions do differ a lot in terms of data collecting efforts and availability of those to the public. One of the key sources of house-specific information is the registration authority data, when a property is registered with the public agency for tax and title purposes. The following list provides some data sources in the UK and US.

CLG Live Tables
CML Statistics
Land Registry HPI
Halifax HPI
Nationwide HPI
US Census Bureau: Housing
FHFA HPI

Case-Shiller HPI
NAHB Housing Economics
NAR Housing Statistics
International House Price Info

In this chapter, I have discussed the economic fundamental of the housing market. One of the key aspects of this analysis is a framework for price determination in the housing market, which is dealt with in detail in the next chapter.

Selected research topics

- The cross-sectional and temporal variation in the depreciation rate.
- Factors influencing housing supply variation across emerging economies.
- Relationship between housing and business cycles.
- Role of unobserved demand shifters in housing demand estimation.

4 House price

Measures and estimation

Chapter outline

- Simple measures
- Hedonic model
- Repeat sales model
- House price index
- Selected research topics

House prices are the most important aspect in the housing market from the perspectives of all economic agents in the market. A consistent estimate of house prices at the national and subnational level can have significant benefits. It provides a benchmark and adds to transparency. However, across many markets in the world, house price measures in the form of a trackable index are not consistently available. In this chapter, we discuss issues related to house price index and methodological frameworks for creating house price index.

Such indices provide transparency in the market. Transparency in the property market is very mixed around the world. Transparency improves with increasing investment activities. Long-term sustainability of businesses and economies requires transparency and a low level of information asymmetry. Market transparency is very important. If we recall the seven attributes that we have discussed in detail in Chapter 2, one of the most important attributes of the housing market is 'information asymmetry'. A transparent market can alleviate information asymmetry. With lower level of asymmetry, the market participants, i.e. buyer and seller as the two most important participants and others such as financiers, investors, policymakers and intermediaries will be able to interact with a higher level of confidence with more public knowledge about market fundamentals. This would greatly facilitate faster and more ready price negotiation, i.e. asking price and offer price being matched to much greater level of satisfaction of all parties and with less time delays. This, at the market level, can help in clearing the stock on sale and improve efficiency. Not only is this beneficial for the investment and business community but also it can create conditions for community wellbeing and inclusiveness.

Table 4.1, compiled from Jones Lang Lasalle (JLL) data, provides a view of transparency relevant to the real estate markets across the world.

JLL's Global Real Estate Transparency Index identifies countries that provide the most favourable operating environments for economic agents such as investors, developers and occupiers. With wide coverage of 109 countries worldwide, it provides a very useful set of information. JLL's Global Real Estate Transparency Index is based on a large number of indicators – 139 variables relating to transaction processes, regulatory frameworks, corporate governance, performance measurement and data availability. A higher real estate

Table 4.1 Global real estate transparency index, 2018

High transparency		Semi-transparency		Low transparency	
UK	1.24			Morocco	3.56
Australia	1.32	China	2.67	Colombia	3.56
US	1.37	Thailand	2.69	Nigeria	3.73
France	1.44	India	2.71	Qatar	3.9
Canada	1.45	Brazil	2.75	Uruguay	3.96
Germany	1.88	Russia	2.78	Iran	3.97
Sweden	1.93	Mexico	2.78	Pakistan	3.99
Transparent		Greece	2.94	Ghana	3.99
Japan	1.98	Croatia	3.01	Panama	4.15
Switzerland	2.02	Botswana	3.06	**Opaque**	
Denmark	2.11	Slovenia	3.06	Lebanon	4.18
Poland	2.15	Serbia	3.11	Honduras	4.5
South Africa	2.21	Philippines	3.11	Iraq	4.51
Portugal	2.3	Chile	3.23	Mozambique	4.53
Taiwan	2.32	Saudi Arabia	3.32	Senegal	4.59
Slovakia	2.4	Peru	3.47	Libya	4.63
Romania	2.49	Argentina	3.47	Venezuela	4.73

Source: Based on data from JLL; for a full list of countries, refer to the JLL report. http://www.jll.com/GRETI

transparency score may be indicative of stronger investor confidence and a higher level of corporate real estate activity.

As Table 4.1 depicts, large number of countries are not transparent. Therefore, improvement is needed across the board. Data collection and processing on market indicators are crucial for that purpose. With greater computing ability and better data capturing tools and processes, we now have a lot more information that can be used effectively for improving transparency. Real estate markets are churning huge amounts of useful data which needs to be brought together to create market intelligence and improve transparency. One of the crucial steps towards better transparency is creation of benchmarks. Benchmarks are useful for several purposes:

1 to understand markets/processes better to generate market intelligence for informing economic decisions;
2 understand behaviour of economic agents;
3 taxation and public goods: mechanisms of efficient taxation;
4 efficient resource allocation – appropriate standards and tools;
5 economic and financial stability: enacting regulation, monitoring loan quality;
6 empowering institutions and strengthening market frameworks: devising monetary and fiscal policies.

House price indices provide such much-needed transparency in the housing market.

Why do we need house price index?

Through market functioning, house prices can capture:

- Physical attributes
- Locational attributes

- Neighbourhood effects
- Local, social and physical infrastructure
- Consumer preferences
- Market trends

It is important to understand house price information to gain key market intelligence as that would aid informed decision-making by various economic agents.

- Information to economic agents for decision-making
- Information asymmetry
- Financing dynamics and loan quality
- Market trends
- Spatial planning policies
- Taxation/fiscal instruments
- Government policies
- Macroeconomic policies

House price index

In order to create house price indices, we need to get back to the economic fundamentals of the market, i.e. demand and supply interactions that lead to market clearing prices or equilibrium values. As discussed in the previous chapter, price expectation plays a crucial role in determining house prices. Therefore, much of consumption and investment preferences of the consumers may be revealed by the choices made in the transaction. Housing is perhaps the most significant and common wealth-creating vehicle for a vast majority of consumers. In a typical human life cycle, most put the majority of their savings in the house that is paid off through mortgage financing. Because of this hugely significant investment, there is an inherent tendency by all homeowners to expect appreciation in the housing capital. Such expectation can take many forms – sometimes it can be the dominating factor and sometimes it may be less so. Regardless, future price expectation can shape a significant part of the current period's housing demand. Price expectation takes more complex patterns during economic uncertainties and fluctuations, e.g. recession and boom period. Much research has been conducted on understanding such patterns. It rarely follows asymmetric patterns. Several theories have been developed to understand the underlying process of formation of price expectation. Two major theories are:

- **Adaptive Expectation** is a hypothesised process by which economic agents form their expectations about the future market dynamics based on past events. For example, if house price appreciation has been higher than expected in the past, people would revise expectations for the future.
    - Prices this period are strongly *related* to prices in previous periods. This lends support to prices being sticky and there is an inherent inertia in economic agents' (buyers and sellers) willingness to revise prices downwards.
    - However, this is irrational. Why should the future be similar to the past? This can lead to forecast inaccuracies and ill-informed investment and policy decisions.
    - However, this is a simple process and can fuel price cycles with sharp changes of direction as prices switch from a rising to a falling regime.

- **Rational Expectation** is an economic theory "holding that investors use all available information about the economy and economic policy in making financial decisions and that they will always act in their best interest" (Merriam-Webster definition).
    - There are two key issues here – (a) economic agents take into account all available information. (b) economic agents will have a good understanding of how the market and economy functions. But, economic agents are rarely so smart! There is weak evidence that people are aware of long-run house price trends.
    - Therefore, economists have postulated weak and strong versions of rational expectations. 'Strong' versions assume actors have access to all available information and always make rational decisions. Any error is due to unforeseen, unpredictable events. 'Weak' versions assume economic agents may not have time to access all information, but they make rational choices given this limited knowledge.

Understanding the psychology behind investment decision-making is the *sine qua non* of issues in behavioural economics. A large number of studies in the last couple of decades made numerous attempts to understand what roles sentiment of economic agents play in shaping up investment decisions and, subsequently, specific market movements through collective or individual actions. As discussed in earlier chapters, sentiment clearly plays a key role in economic performances, but its impact under various economic environments is not straightforward as it may change for different types of agents at different times of the economic cycle. Moreover, causality effects are also not clear as market economies may be both driven and shaped by the agents' behaviour.

The related theoretical frameworks offer explanations through the presence of 'animal spirits', possibility of 'habit persistence' and forward-looking models. While there is no doubt about the existence of this effect, its extent is rather unclear despite several attempts to estimate it using different tools and models. Therefore, a key question is: how precisely can the investors' attitude provide us important indications for future market imbalances?

The real estate sector provides some unique features that are in line with the twin motives of market imperfections and precautionary savings behind the rational expectations permanent income hypothesis (Hall, 1978). These features are (see Marcato and Nanda, 2016):

- Consumption is typically disproportionate. This is due to the fact that real estate purchase decisions are infrequent and investment tends to be lumpy. This causes spikes in the patterns of investments.
- Real estate investors try to hold cash as much as possible because they can use the cash holding in times of quick redemptions or if they need to pay off loans when property prices are falling. Along with this, the distributed lag of past actual income can reflect a timing mismatch between the time the investment decisions are made and the actual contracts are exchanged money is transferred (6–9 months trading period).
- As a result of the above behaviour, the precautionary saving is a common and an important motivation for holding assets to meet long-term loan obligations and investment commitments.

In the housing market, we have seen evidence of asymmetric price expectations repeatedly. Such behaviour often result in significant price cycles. A multitude of factors determine the variation on the demand and supply sides and follow intricate, long and variable lags; overlapping effects. Shocks to many of these factors can trigger upturns or downturns, e.g. credit availability, employment. It is hard to forecast changes, especially turning points, because of this complexity. But, economic agents still do forecasting! Most of the demand variations are transmitted through price when supply is inelastic, especially in the short-run and to some extent even in the long-run, e.g. the UK. Stock overhangs can aggravate downturns, e.g. the US, Spain and Ireland in 2007–2010.

Hedonic model

The theoretical foundations and empirical framework of the hedonic model have been developed through many works (see Rosen, 1974; Edlefsen, 1981; Roback, 1982; Bartik, 1987; Epple, 1987; Blomquist et al., 1988; Bajari and Benkard, 2005 and many others (see review of literature by Malpezzi, 2003). Essentially, the capitalisation process works through constrained utility maximisation by the consumers. Key points from a Rosen (1974) paper are:

- It is plausible to view economic goods as a bundle of attributes.
- The values of the attributes of an 'economic good' get capitalised into observable prices thorough the Revealed Preference Approach, i.e. the implicit price of a multi-faceted good. Households receive utility from each of the many attributes of the house and therefore they are willing to pay for the attributes. It is also plausible that there may be disutility from the undesirable attributes which may be subjective to specific households based on their preference sets.
- In equilibrium, the marginal price = marginal willingness to pay for an additional unit.
- In an empirical set-up, we estimate the value of an additional quantity of an attribute, holding all other attributes constant (*ceteris paribus*).

The Hedonic framework allows us to answer several important and relevant questions, such as:

- How much does an additional bedroom (or bathroom or any space) add to the sales price of a house? What is the value of hardwood floor? Or marble countertops?
- Impact of school quality on house prices: do higher average student test scores increase the value of houses within the school district?
- What is the impact of air quality on house prices?
- What is in 'a view'? How much is the value of the 'water front' view for a property?
- Do higher crime rates reduce house prices?
- What are the influences of rock quarrying activities on home and small business roperty values?
- And so on…

In a further refinement and development of the Rosen framework, Roback (1982) incorporated labour market as often housing market and labour market outcomes are jointly determined by the economic agents with much dependence on each other. The Rosen model may work well in a single metro area – where we can perhaps assume that any difference in the attributes will be capitalised into prices and labour market is roughly uniform. However, across metro areas, we need to consider the labour market dynamics. For example, households may be willing to accept a lower level of air quality with a higher level of wages. The Roback model combines housing and labour markets. A combination of wages and land rent gives levels of profit for firms and utility for households.

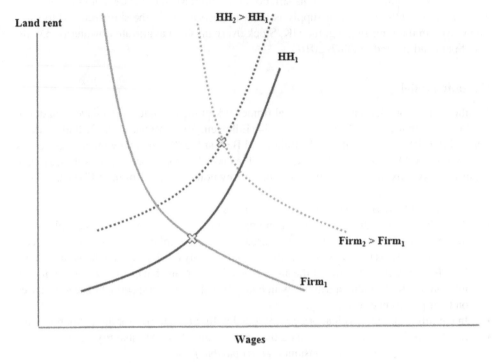

Figure 4.1 Households and firms equilibrium, combining labour market dynamics.

Source: Adapted from McDonald and McMillen, 2011, see pp. 210–212 for a full analysis.

In Figure 4.1, households' preferences are shown in terms of land rent and salary wage combinations, which is a positively sloped curve implying that a household is willing to pay higher rent with higher level of salary income and vice versa. For the firms, it is negatively sloped as firms will be less willing to pay a higher wage if the land rent is high. Hedonic equilibrium is reached when a firm's offer and household's bid are equal. We can derive a hedonic price function through determination of such equilibrium combinations for all household and firms.

Empirical aspects of hedonic framework[1]

The starting point in hedonic analysis is to simplify the problem with some reasonable assumptions. Housing can be justifiably assumed as a differentiated commodity, which is characterised by many features, facets and attributes, i.e. a vector Ψ of various attributes that can be associated with the physical features of the unit and the location of the house.

The examples of the components that can be part of Ψ are: physical attributes [age of the unit in years (y), size of interior built area of the unit square footage (sf), number of bedrooms (b), number of floors (f), presence of a garden (g_1), presence of garage (g_2)] and locational attributes [school quality (sq), crime rate (cr), transport link (tr), or a composite metric for locational desirability of the unit (L)]. Transport infrastructure plays a very significant role in shaping up the property market. Research shows significant investment effect in the property

---

1 This discussion is drawn from the empirical framework presented in Fuerst et al. (2015, 2016).

market due to infrastructure development. The infrastructure projects open up the land and creates opportunities for further development. The evidence is strong in both developed and developing countries.[2] For example, Ball and Nanda (2014) investigated whether increases in infrastructure investment across the regions in England led to boosts in development of new commercial and residential property. Using time series econometric models, authors analysed both physical (roads and harbours) and social infrastructure (education and health) impacts. While the results were mixed and somewhat inconclusive at the aggregate level, the implications of the infrastructure investments at the local level can be significant and powerful. Nanda and Tiwari (2013) also found strong spillover effects of infrastructure in the Indian city of Bengaluru. Across the countries and cities, the effects do depend on the existing infrastructure endowment. In developing countries or places with low levels of infrastructure, the effect can be much stronger compared to those in the developed countries. Anyhow, the key point is that the transport links are a very important aspect of locational desirability.

Locational attributes may be included by specifying the granular components that make up the assessment whether a location is desirable or not from a household's perspective i.e. can be assigned a composite score. Distance from places of economic and social interests play a key role and is perhaps the most defining aspect of housing location, as a household would like to have easy access and proximity to those places of interests. Moreover, distance may involve various costs of transport – time, money and psychological cost of undertaking the travel. Therefore, we can include a distance term (d) from a town centre or village centre as a variable indicating desirability of the location.

One of the key controls needed in this framework is the overall quality indicator for the unit. Imagine a unit with a better quality of material with the latest designs of the fixtures and fittings. So, we can also include a variable indicating quality of the housing unit, hq. Note that this will be tricky to determine as it can be a bundle of various factors and a complex combination of many other attributes. For the purpose of deriving the utility function, we assume one undifferentiated numeraire commodity, K. K is a measure for all other non-housing goods and services that a household may consume during the consumption period.

The utility function, having the usual continuity and convexity properties (which are not tenable for non-continuous variables), is then:

$$U = U(\Psi, K) \tag{4.1}$$

$$\Psi = \Psi(y, sf, b, f, g_1, g_2, sq, cr, tr, d, hq) \tag{4.2}$$

As mentioned before, housing quality, hq, is a tricky variable to measure – it can have many aspects and it can also differ depending on a household's perspectives towards and evaluations of those quality features. A household may care more about the kitchen quality than another household. However, this can, to some extent, be sorted by the market process. hq may be a combination of quality of the building material (bq), architecture and design aspects (dq) and quality of utility services, i.e. running water, electricity, heating or air-conditioning (uq), etc.

---

2 "Does Infrastructure Investment Stimulate Building Supply? Case of the English Regions" 2014 (Ball, M., Nanda, A.) *Regional Studies* 48(3): 425–438. http://www.tandfonline.com/doi/abs/10.1080/00343404.2013.766321
  "Sectoral and Spatial Spillover Effects of Infrastructure Investment: A Case Study of Bengaluru, India" 2013 (Nanda, A., Tiwari, P.) Published by The Royal Institution of Chartered Surveyors (RICS). http://www.isurv.com/site/scripts/download_info.aspx?fileID=5213

$$hq = hq(bq, dq, uq) \tag{4.3}$$

So, equation (4.1) can now be written as:

$$U = U(y, sf, b, f, g_1, g_2, sq, cr, tr, d, bq, dq, uq, K) \tag{4.4}$$

In this set-up, one can ask questions on several aspects and attributes, i.e. how much is a consumer willing to pay for a house with a specific attribute, e.g. a bedroom? Or, how much value would a typical household place on having a garden or a garage? Also, note that the specification in equation (4.4) can be made much more complex with finer levels and categories of variables and more granular definitions. I point out the estimation issues related to that later in this section.

If we now look at the consumer's problem, a budget constraint needs to be specified. A household's budget is constrained and limited over a time period, e.g. monthly salary or income. A household makes choices for housing, non-housing and savings each month from that salary or income. This is a household's utility maximisation problem. Obviously, the budget constraint plays a very important role as his/her consumption level (and utility derived from that consumption) depends on the income level.

Let's refine the notations. For a level of salary or income I_0, the consumption basket is C_0 and derived or achievable utility from that consumption basket is given as U_0. The price of K (non-housing composite), k, can be normalised to unity and the bid-rent for the house is R. A household may wish to save a portion of the income each month. That savings s may earn some interest at the rate of r per cent. r can be uncertain as interest rate changes due to monetary policies and market conditions. So, interest income from savings can be associated with a probability distribution. So the individual's budget constraint can be written as:

$$R + k + s = I + rs \tag{4.5}$$

rs is the interest income from savings s.

The *bid-rent*, R, is the consumer's willingness to pay for Ψ. Therefore, equation (4.1) may be represented by:

$$U = U(\Psi, K) = U\{y, sf, b, f, g_1, g_2, sq, cr, tr, d, bq, dq, uq, (I + rs - R - s)\} \tag{4.6}$$

Inverting equation (4.6) leads to a *bid-rent* function as follows,

$$R = R\{y, sf, b, f, g_1, g_2, sq, cr, tr, d, bq, dq, uq, (I + rs - s)\} \tag{4.7}$$

Using any attribute of $h(.)$, the shape of the bid-rent function in equation (4.5) is then represented by the partial derivatives, e.g. with respect to number of bedrooms or school quality or crime rate.

$$R_{sq} = U_{sq} / U_K \tag{4.8}$$

Let's take an important physical attribute that a household looks for in choosing a house: number of bedrooms.

Figure 4.2 shows that there are increasing bids at a decreasing rate. There are maximum amounts that a household would pay for an additional unit and there are increasing offers at increasing rates from the firms' point of view.

Figure 4.3 shows the points of hedonic equilibrium (left panel) where bid and offer curves interact. In the diagram, we show examples of two such points where a household's bid is equal to a firm's offer amount. If we have n number of households, we can find such hedonic equilibrium points for all such combinations of bid and offer values and connecting those equilibrium points would yield a hedonic price function (right panel) for the sample of properties that we choose to analyse. In an empirical set-up, it is the hedonic price function that we attempt to determine for all the relevant attributes. Let's now have a closer look at how the empirics may work.

Note that equation (4.8) is the maximising condition of price being equal to marginal rate of substitution. The assumption of a strictly diminishing marginal rate of substitution implies that bid rent R will increase with an attribute at a decreasing rate. Therefore, the slope of the bid rent function can be interpreted as a household's willingness to pay for an additional unit of an attribute, which is called the hedonic price of a particular feature.

Hedonic studies typically use equation (4.8) within a standard regression technique such as Ordinary Least Squares (OLS) and the focus is on the price effect or the coefficient of each parameter, i.e. to derive equation (4.8). However, utmost care is needed while estimating the regression equation. As already indicated on several occasions, equations (4.1–4.8) may look simple but the challenge is in finer specification of the variables and how we recognise and control their inter-relationships. This can easily render a regression estimation useless due to the biases.

As Fuerst et al. (2015, 2016) note, the sources of biases are several, as mentioned below.

First, the physical features of a house can be inter-linked among themselves. Key physical and locational features tend to be associated with the type and nature of all other factors, e.g. a house with 3–4 bedrooms will likely to have more than 1–2 bathrooms and will have at least a level of square footage that can support these rooms. This implies the presence of correlation among the variables in equation (4.7), which is referred to as *multi-collinearity* in econometrics.

Second, as mentioned, the quality variables for the housing unit and the location and the judgement on those from a household's perspective can be fraught with many unobservables

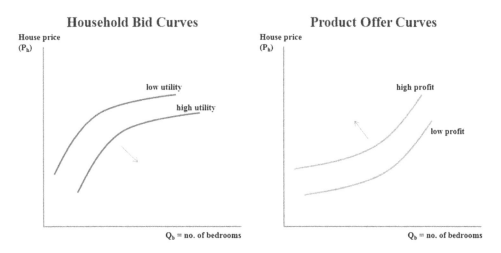

Figure 4.2 Household bid and product offer curves.

Source: Adapted from McDonald and McMillen, 2011, see pp. 206–208 for a full analysis.

48 *House price*

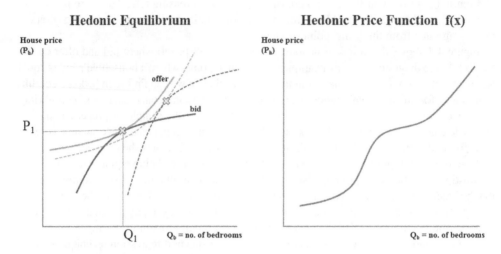

Figure 4.3 Hedonic equilibrium and price function.
Source: Adapted from McDonald and McMillen, 2011, see pp. 206–208 for a full analysis.

that may not be captured in standard data collection effort or easily controlled for. This is called *omitted variable bias*. Typically, such issues are dealt with *unit* and *location fixed effects* in econometric specification.

Third, the savings and alternative investment opportunities come with significant uncertainties due to changing market conditions. Therefore, the opportunity costs from a household's perspective may change over time.

Finally, although equations (4.1–4.8) appeared linear in the specification, some unit and location attributes will need to be non-linear for correct specifications. For example, the price of a housing unit might respond to age non-linearly and also to the size of the property due to higher maintenance costs and under-consumption possibilities. An old building with historical design value may fetch a significant premium.

Next, we turn to the empirical estimation strategy appropriate for addressing a number of issues raised above.

Hedonic estimation strategy

Equation (4.7) can be estimated empirically with all the required data on the feature. Readers are encouraged to refer to an econometric book for exposition on the regression techniques. Typically the regression specification is as follows:

$$HP_{ijt} = \propto_{it} + \sum_{i=1}^{I} \beta_i X_{ijt} + v_{ijt} \qquad (4.9)$$

where HP_{ijt} is the transaction price of a property i in location j in time t (typically specified as the natural logarithm of price in currency unit per size unit), X_{ijt} is a vector of observable explanatory locational and housing unit variables; β_i is a vector of coefficients to be estimated; and v_{it} is the disturbance term which is assumed to follow a normal distribution with a mean of zero and constant variance. As mentioned before, to capture the location (C_j) and time fixed effects (Y_t), dummy variables are also included:

$$HP_{ijt} = \alpha_{it} + \sum_{i=1}^{I} \beta_i X_{ijt} + Y_t + C_j + v_{ijt} \tag{4.10}$$

Fuerst et al. (2015, 2016) analysed the effect of energy efficiency features in the housing market using the above modelling specification.

Let's look at another example of hedonic application with significant research interests. Many studies have looked at the effect of school quality on house prices. Clapp et al. (2008) used a sample of sales of owner-occupied properties with one to three units spanning over 11 years from 1994 to 2004 across towns in the state of Connecticut, US. The sample has more than 350,000 transactions for one to four unit owner-occupied structures, after all filters to eliminate invalid or non-representative transactions. The paper examines the effect of school district performance, as measured by student test scores, and the effect of the student socio-economic and demographic composition on local property values.

The dataset contains information about:

- the unit address
- selling price
- assessed value
- sales date

The detailed listing of the unit characteristics such as:

- internal square footage;
- number of rooms;
- number of bedrooms;
- number of bathrooms;
- age of the property;
- size of the parcel of land;
- and, a number of neighbourhood attributes such as school quality, owner-occupancy level in the neighbourhood, level of poverty, etc.

The results confirm the importance of school quality in house prices. Table 4.2 shows results for some standard hedonic attributes from Clapp et al. (2008).

Table 4.2 Example of hedonic estimation results

Control variables	OLS estimates
Number of rooms	0.240 (10.07)
Number of bedrooms	0.015 (4.75)
Age in 100s	−0.817 (−20.69)
Log(square footage <2500 sq.ft.)	0.480 (32.97)
Log(square footage >=2500 sq.ft.)	0.076 (2.40)

The dependent variable is the natural log of house price. t-stats are in parentheses. Model includes several other variables. Refer to the full table.
Source: Adapted from Table 3 from Clapp et al. (2008); for a full analysis, refer to the source.

In the hedonic framework that I have discussed so far, a key missing element is effect of the neighbouring housing units. How do neighbour's housing characteristics affect someone's housing consumption and choices? A spatial hedonic framework incorporates the characteristics of the neighbouring housing unit. The closer the neighbouring unit is the stronger may be the effect. However, it also depends on the matching between the properties. In an application of spatial hedonic framework, Nanda and Yeh (2016) studied the conspicuous consumption in terms of whether such consumption can lead to positive or negative net externalities.[3] The authors examined how the presence of luxury and expensive properties within a local area affected the house prices in Taipei City. While positive externalities can be found through improved urban amenities or *reflected glory*, the negative externalities can be associated with the adverse reference group effect; that is, *repulsive envy*. In the Taipei City case study, the authors found significant spatial patterns in house prices and the importance of socio-economic influences on house prices. The methodology is based on the spatial hedonic framework as described below.

Following on from equation (4.10), the neighbouring unit's attributes can be incorporated as:

$$HP_{ijt} = \rho W HP_{ijt} + \sum_{i=1}^{I} \beta_i X_{ijt} + Y_t + C_j + v_{ijt} \qquad (4.11)$$

where ρ is a spatial autocorrelation parameter used to measure the spatial interaction of house prices; W is an $n \times n$ spatial weight matrix which is predetermined by contiguity with n observations. The value of spatial correlation is 1 if the observation i and observation j are neighbouring units. If they are not neighbours, the value is 0. The spatial matrix is normalised with each row summing to unity. However, it is also conceivable that not only house price but also the specific attributes of the neighbouring unit may matter. This can also be incorporated in the above framework, equation (4.11), by including attributes, e.g. number of bedrooms or size of the neighbouring property.

Price index example

Several house price indices are derived using a hedonic framework. The method uses detailed property characteristics to create a benchmark transaction and the benchmark is then used to calculate price differences for all other transactions in the sample. For example, the house price index can be derived from information on the following house characteristics:

- purchase price;
- location (region, city/town);
- type of property: house, sub-classified according to whether detached, semi-detached or terraced, bungalow, flat;
- age of the property;
- tenure: freehold, leasehold, feudal;
- number of rooms: bedrooms, living rooms, bathrooms or size of the unit in square foot or square meter;
- number of separate toilets or bathrooms;
- central heating or air-conditioning;

---

3 "Reflected Glory vs. Repulsive Envy: What Do the Smiths Feel about the House of the Joneses?" 2016 (Nanda, A., Yeh, J.) *Asian Economic Journal* 30(3) 317–341. http://onlinelibrary.wiley.com/doi/10.1111/asej.12095/full

- number of garages and garage spaces;
- garden: types of garden, size of garden;
- land area;
- Etc.

Repeat sales model

The Hedonic framework discussed so far often suffers from a couple of major problems:

- there is an *omitted variable bias – some attributes are left out from the data and are not observable*;
- the assumption of *no structural change is unrealistic*.

As a result of these two problems, hedonic estimation ends up comparing properties which are not matched properly. This poor matching issue gets exacerbated if the dataset contains poor quality of property and locational attributes. Therefore, the repeat-sales method is an effective way of calculating changes in the sales price of the same housing unit over specific periods of time in the sample. Effectively, the problem of poor matching is taken care of by comparing two transactions of the same house. Various housing price indices use the repeat-sales method and are frequently used for understanding the housing market by buyers and sellers, investors, banks and policymakers.

House price increases may reflect a combination of land price appreciation and the structure price appreciation. Repeat sales indices try to difference out 'structure' fixed effects – isolating the effect of changing land prices. One of the key assumptions made in the repeat sales framework is that the structure remained constant between two transactions. So, the repeat sales model:

- compares 'like' with 'like'
- assumes constant quality over time

The repeat sales method uses data on properties that have sold at least twice, in order to capture the true appreciated value of constant-quality houses. The main variable is the change between two transaction prices of the same unit.

An example

- Repeat Sales: A home in London was sold in 2000 for £300,000, and it was sold again in 2010 for £450,000, i.e. 50% appreciation.
- In the house price index calculation, the 50% appreciation over 10 years will be compared with information within a sample of such repeat sales transactions to calculate how much the same and ***constant quality*** home had appreciated.

It is important to note that while several common factors contribute to house prices (and house price index construction) in all countries, there can be country or location specific factors that should be duly considered in the index formulation. For example, Kavarnou and Nanda (2015) analysed the Panamanian context and created a house price index.[4] They

---

4 "House Price Dynamics in Panama City" 2015 (Kavarnou, D., Nanda, A.) *Journal of Real Estate Literature* 23(2) 315–334. http://aresjournals.org/doi/abs/10.5555/0927-7544.23.2.315

outline the unique characteristics of the market. Using a dataset of property-level information over 2007–2014, the authors employ a hedonic modelling framework to analyse the impacts of certain amenities and drivers that may affect housing values. The results indicated several unique features of the Panamanian housing market. Also, for example, in another study, Kavarnou and Nanda (forthcoming) analysed the importance of the tourism sector in the housing market.[5] Using Crete Island as a case study and data from 2006 to 2012, the authors constructed tourism indicators for the four prefectures of Crete, which are then used to create a tourism penetration rate (TPR) for each prefecture. The study then performed a hedonic house price regression analysis to establish that (i) the TPR of a prefecture has a significant effect on house prices of the prefecture and (ii) house prices are affected by the TPR of the neighbouring prefectures, indicating statistically significant tourism spillover effects.

One of the most prominent house price indices that uses the repeat-sales method is the Case-Shiller Index. The underlying method does not take into account new construction, condos and co-ops and any non-arms-length transactions, such as home sales between family members at below-market prices. However, foreclosure sales are included in the sample. Other indices that use the repeat-sales method are the Federal Housing Finance Agency's (FHFA) monthly House Price Index, which is based on Fannie Mae and Freddie Mac's data on single-family home sale prices and refinance appraisals; and CoreLogic's Loan Performance Home Price Index. Canada's major home price index, the National Composite House Price Index, also uses the repeat-sales method. In the UK, the Land Registry House Price Index is based on the repeat sales methodology. However, there are two significant problems with the repeat sales method:

- Sample selection bias: it is possible to end up with a sample of high frequency transacting houses for repeat sales analysis. This will be affected by time period and economic cycles. First-time transactions are not included, which can exclude a significant part of the market.
- The assumption that there are no changes in property characteristics and their parameters between two transaction dates is unrealistic. Often transactions take place over more than 5 years and during this time many owners may choose to alter or add a significant part to the house that can carry sizeable price effects. At the least, maintenance expenditure, to preserve the capital value and delay obsolescence, can be significant over a long period of time.

Despite these drawbacks, the repeat sales technique still offers a reasonable framework for estimating house price indices. Let's look at a selection of frequently used house price indices in two major economies – the UK and US.

Types of house price measures

There are several house price measures usually available by various government and non-governmental agencies. Those are:

- Simple average house prices
- Weighted average house prices

---

5 "How Does Tourism Penetration Affect House Prices? Evidence from Crete, Greece" (Kavarnou, D., Nanda, A.) Forthcoming at *Tourism Analysis: An Interdisciplinary Journal*.

- Median house prices
- Hedonic house price index
- Repeat sales house price index

In the UK, some of the commonly used indices are:

- Communities and Local Government (CLG) simple average prices
- CLG Mixed-Adjusted House Price Index (weighted average)
- Halifax House Price Index (based on Hedonic method)
- Nationwide House Price Index (based on Hedonic method)
- Land Registry House Price Index (based on therepeat sales method)

Tables 4.3 and 4.4 summarise various commonly used indices in the US and UK.

In the United States, frequently used house price measures are:

- US Census Bureau Average and Median New Home Prices
- National Association of Realtors® Median Sales Price Series
- FHFA House Price Index (formerly OFHEO)
- S&P/Case-Shiller Home Price Index
- Loan Performance House Price Index
- Others:
  - Radar Logic House Prices per Square Foot
  - Constant Quality (Laspeyres) Price Index of New One-Family Houses Sold
  - Zillow Home Value Index

Indices of repeat sales are more commonly used and regarded as better than other indices such as averages and medians as those control for quality of houses. In Table 4.4, I summarise two types of house prices indices in the US. The comparative analysis shows differences in the respective samples and methodologies.

As discussed so far, due to difference in methods and sample, house price measures differ significantly, especially during certain periods. Let's look at a selection of house price measures from the US and UK and plot the values over a sufficiently long time period to cover at least a couple of full economic cycles.

Figure 4.4 shows per cent annual change in two commonly used house price measures in the UK from 1996 to 2010. These measures are: Land Registry Index based on repeat sales

Table 4.3 UK house price indices

	Type	Method	Data/sample
Halifax	Sold price index	Hedonic regression	Mortgage approvals
Nationwide	Sold price index	Hedonic regression	Mortgage approvals
ONS	Sold price index	Hedonic regression	Mortgage completions sample
Land Registry	Repeat-sales index	Repeat-sales regression (RSR)	Registered transactions, 100% of sales England and Wales
LSL Acad	All registered sales	Mixed adjustment	Land registry price paid data
Rightmove	Asking price index	Mix adjusted	Asking price listing on Rightmove portal

Source: various.

Table 4.4 US house price indices: A comparison

Median indices	Repeat sales indices
• The National Association of Realtors (NAR) index has been tracking house prices since 1968 and it is constructed as a median index. • The source of data is the survey of sales of existing single-family houses from NAR affiliates. • The national median is computed by value-weighting the median by the number of units in the sample. This is calculated for all four census regions in the US. • The Census Bureau house price index is also calculated as a median index. However, unlike the NAR index, the sample includes new units. • Since new houses tend to be priced higher than the existing house, the Census index often reports a value higher than the NAR index.	Three frequently used repeat sales indices in the US are: • The Federal Housing Finance Agency (FHFA) index • The S&P/Case-Shiller index • The CoreLogic Index • Since 1975, the FHFA index has been published quarterly covering a large sample of properties. The sample of units is obtained from mortgages securitised or processed by Fannie Mae or Freddie Mac only. • Apart from having units processed by Fannie Mae and Freddie Mac, another very important distinctive feature of the FHFA index compared with the other two indices mentioned above is that the sample includes refinance mortgages. This is not the case with the Case-Shiller and CoreLogic indices. However, FHFA also publishes purchase-only versions of the index. • The Case-Shiller and CoreLogic indices include all transactions on single-family homes, including those which are not prime-market products. The samples tend to have a significant share of volatile market units which transact more often than the average. • The Case-Shiller methodology includes an interval-weighting procedure with higher weight placed on short-internal repeat sales (re-transacted within a short time period). CoreLogic index does not employ such weighting. • CoreLogic has larger coverage than Case-Shiller as it includes a higher number of US states.

Source: Based on information from the NAR, US Census Bureau, FHFA, Fiserv, CoreLogic.

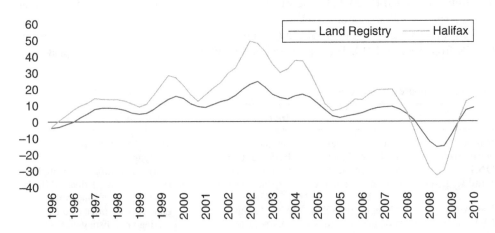

Figure 4.4 UK house price: Comparison by methods (% year-over-year growth).

Source: Based on data from Lloyds Banking Group plc, Land Registry.

method and Halifax based on the hedonic framework. As reflected by the chart, while the long-run trend is similar for all four indices, there are time periods when the price indices differed markedly and often provided very contrasting results.

Figure 4.5 provides a similar exercise for the US house price measures.

In Figure 4.5, two commonly used indices based on different methods are plotted over a long time period (1988–2009). As in the case of the UK figure, it also shows similar long-run trends reflected by different measures while showing a remarkable difference at certain time periods. These two examples beg the question: **what should we use for all purposes including investment decisions?**

It depends on the context and purpose, e.g. median is useful for quick comparison but we need to be careful about the sample. It is not advisable to mix different types of houses (e.g. bungalows and flats) as the demand and supply dynamics are quite different across various house types. Moreover, houses as economic goods are heterogeneous. House price index methods vary, and all have merits and demerits. However, it is better to use the one that compares similar units, i.e. 'apples with apples'. The repeat sales indices measure house price appreciation better compared to other methods. We should consider the context and understand the underlying data and time period. It is always advisable to compare price fluctuations in a 'normal' market with relatively less severe demand-supply mismatch. *Overall, we should infer on the house price situation based on a combination of measures.*

House prices are sensitive to interest rate changes. Increasing reliance on the mortgage market to finance a house purchase has made households more prone to financial fluctuation due to interest rate changes. A small amount of change such as 25 or 50 basis points can lead to a significant increase in monthly mortgage payments. Such reliance has deepened over the last three decades with much developments in the mortgage financing market, which have experienced many innovative products and services as well as influx of investment money for the financial institutions. Historically, it has been shown that the sensitivity or elasticity (house prices to 100bps increase in short-term interest rate) has changed from 1970 to 1982 and 1983 to 2007 for a number of countries (IMF estimates).

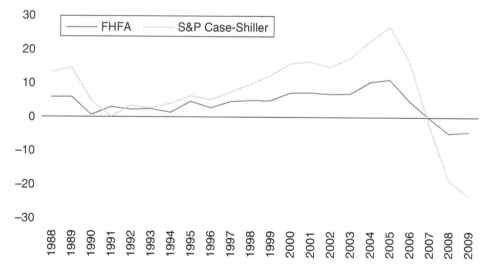

Figure 4.5 US house price: Comparison by methods (% year-over-year growth).

Source: Based on data from FHFA, Fiserv.

56 *House price*

We also find significant divergence of house prices due to fundamental demand shifters and non-fundamental factors. Changes in population (size and composition), employment, income and interest rates (cost of financing) are the most significant fundamental demand shifters. Most of the house price fluctuations can be and should be driven by these fundamental factors. Yet, across many countries and at certain times, we have experienced significant contributions of the non-fundamental factors, mainly driving the speculative housing demand. IMF estimates show a sizeable amount of per cent increase in house prices during 1997–2007 that could not be explained by fundamental drivers of house prices. Several countries have experienced such divergences, which have often resulted in sharp increase and decreases in house prices leading to economic and financial crises such as in Ireland, Netherlands, the UK, etc.

A very interesting phenomenon is often experienced across many parts of the world which is called the 'Ripple Effect'. Several countries have been studied for example, Lee and Chien (2011), Payne (2012), Balcilar et al. (2013) and Lean and Smyth (2013) consider the existence of a ripple effect for Taiwan, the US, South Africa and Malaysia respectively. The UK regional housing markets have been studied by multiple authors – Holmans (1990), MacDonald and Taylor (1993), Alexander and Barrow (1994), Drake (1995), Ashworth and Parker (1997), Meen (1999), Petersen et al. (2002), Cook (2003, 2006, 2012), Cook and Thomas (2003), Cook and Watson (2016), Holmes and Grimes (2008) and Holmes (2007). These studies consider the possibility that house price changes are first observed in London before moving to other regions of the UK, depending on spatial linkages.

The results have been mixed with some evidence supporting the ripple effect, i.e. showing a positive relationship between the geographical proximity of a region and the degree of the changes in house prices. Several factors can contribute to such phenomenon-proximity, economic linkages and dependence, people's commuting pattern over multiple regions,

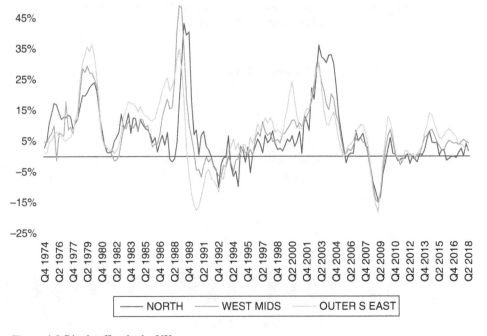

Figure 4.6 Ripple effect in the UK.

Source: Based on data from Nationwide House Price Index.

House price 57

alignment of regional policies, etc. Figure 4.6 shows changes in Halifax house price index across three regions in England – North, West Midlands (WM) and Outer South East (SE). It is evident that Outer SE is leading the way over the time period shown, which is quite prominent in some time periods.

Global house prices

Globally, house prices have shown much variation across major economies. Figure 4.7 shows the global house price index over the last several years. Several well-known cycles are evident, along with a recent run-up in price levels. It shows the dramatic uptick in price level prior to the 2008 Global Financial Crisis (GFC) and the slowdown afterwards.

Several country-level factors contribute to the variation. The International Monetary Fund (IMF) provides a summary of house price levels across a number of countries. Figure 4.8 shows real house price annual per cent changes over the past year (2017Q2 or the latest). Several countries with political and economic turbulence showed falling house prices, e.g. Ukraine, Russia, Qatar, Greece, Vietnam, etc. A large number of countries showed much growth in house prices with more than double-digit growth in Hong Kong and Iceland.

Much of the house price inflation across the countries can be attributed to credit growth. Figure 4.9 shows real credit growth over the past year in 2017Q3 or latest available in annual per cent change. Hong Kong and Philippines showed much growth along with Mexico.

Mortgage markets across many developing countries have gone through much development and the popularity of mortgage financing as a source of fulfilling homeownership aspiration has fuelled the demand and supply of mortgage credit in these markets.

Along with house price inflation comes the affordability concerns. One of the simplest and most effective way of determining affordability is to take the ratio of income and house prices. Typically, average, median or lower quartile house prices and income levels are used

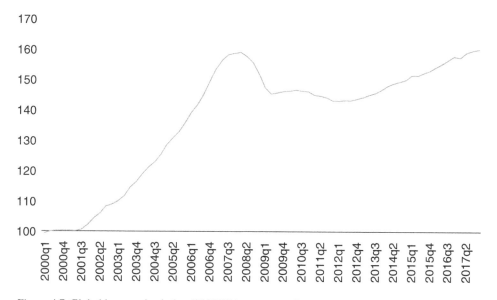

Figure 4.7 Global house price index (2017/18 latest quarter).

Source: Based on data from IMF. http://www.imf.org/external/research/housing/

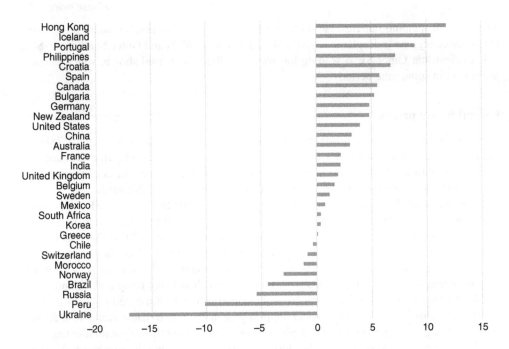

Figure 4.8 Global house price growth (2017/18 latest quarter).

Source: Based on data from IMF; for a full analysis, refer to the *source*. http://www.imf.org/external/research/housing/

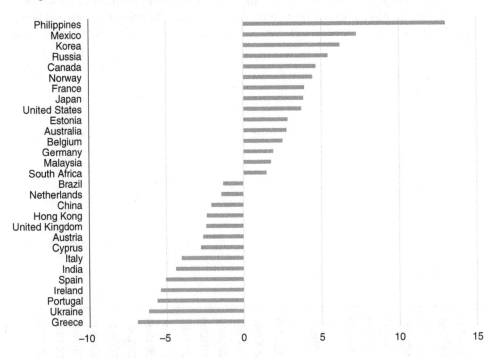

Figure 4.9 Global credit growth (2017/18 latest quarter).

Source: IMF. http://www.imf.org/external/research/housing/

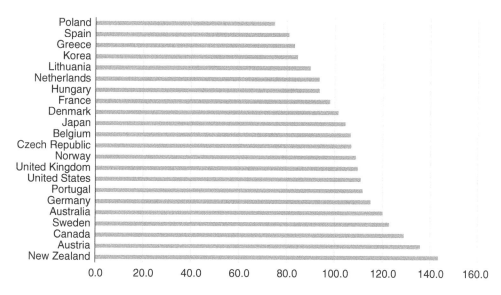

Figure 4.10 Global housing price-to-income ratio (2017/18 latest quarter).

Source: Based on data from IMF; for a full analysis, refer to the *source*. http://www.imf.org/external/research/housing/

to calculate the affordability ratio. In England, financial institutions use an income multiple of about 4.0–4.5 as an indicator of an applicant's affordability. Figure 4.10 shows that the house prices have grown faster than incomes across a number of countries including some major economies. The widening gap between house price growth and income growth is a key policy concern. A sustained level of widening gap over a long time period can create much social divergence and further the rich-poor divide.

An increasing level of divergence of house prices and rent can significantly drive changes in tenure choice in the local housing market. According to the user cost of housing framework, a comparison between monthly housing costs (including monetary and non-monetary items) of ownership and renting can drive relative changes in home purchase and renting demand. Figure 4.11 shows house price appreciation outpacing significantly across a large number of countries with much implications for the housing sub-sectors.

Research shows that global housing markets may be linked. The movement and effects of house prices in one region can be linked with those in other regions. Economic connections among the economies are the key regions. Especially with a much more integrated global financial market and the intense level of international trade over the last several decades have made the inter-linkages among the countries quite strong. The inter-linkages of housing markets among the countries within trade blocs or geographic regions can be especially strong. For example, Nanda and Yeh (2016) analysed the international transmission mechanism and its role in contagion effect in the housing markets across six major Asian cities.[6] The analysis is based on the identification of house price diffusion effects using quarterly data for six major Asian cities (Hong Kong, Tokyo, Seoul, Singapore, Taipei and Bangkok) from 1991Q1

---

6 "International Transmission Mechanisms and Contagion in Housing Markets" 2016 (Nanda, A., Yeh, J.) *World Economy* 39(7) 1005–1024. http://onlinelibrary.wiley.com/doi/10.1111/twec.12288/abstract

60 *House price*

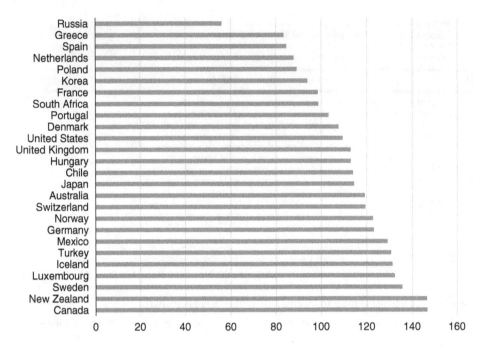

Figure 4.11 Global house price and rent ratios (2017/18 latest quarter).

Source: Based on data from IMF; for a full analysis, refer to the *source*. http://www.imf.org/external/research/housing/

to 2011Q2. The empirical results show that the open economies (those which depend on international trade, e.g. Singapore, Japan, Taiwan and Thailand) have close and positive association between the economy's openness and house prices. However, authors also note that some region-specific conditions can influence the relationship significantly, e.g. restrictive housing policies and demand–supply imbalances such as in Singapore and Bangkok. Such results and insights are very useful for devising investment strategy and public policies in the housing market.

In this chapter, I have discussed the house price determination and highlighted major challenges in measuring house prices. One of the key challenges that I have noted is the difficulty in capturing locational aspects fully into the price models such as the hedonic framework. In the next chapter, I deal with location choice issues with some theories and reference to practical issues.

Selected research topics

- Unobserved factors in house price estimation and role of new forms of data.
- Global variation in the underlying drivers of house prices.
- Contagion effect in global housing market.
- Approaches of solving global housing affordability problem.

5 Location choice in urban areas

Chapter outline

- Facts and figures of urbanisation
- Location choice, models of urban areas
- Land use planning
- Housing and neighbourhood
- Selected research topics

Urban areas evolve over time through firms, households and other economic entities deciding to locate due to their economic interests. One of the key attributes discussed in Chapter 2 is immovability or spatial fixity. This chapter presents the theories and emerging trends in this respect. Our focus is on understanding the models of location choice and how housing consideration are key to formation and development of urban areas. First, let's look at the scale and nature of urbanisation globally.

Facts and figures of global urbanisation[1]

Following are the facts and figures on urbanisation compiled by the United Nations.

- In 1800, only 2% of the world's population was urbanised. In 1950, only 30% of the world's population was urban. In 2000, 47% of the world's population was urban. In 2016, an estimated 54.5% of the world's population lived in urban settlements. By 2030, it is expected that 60% of the world's population will live in urban areas. Almost 180,000 people are added to the urban population each day. It is estimated that there are almost a billion poor people in the world; of this over 750 million live in urban areas without adequate shelter and basic services. This creates huge needs for effective policies to tackle the problem of urban poverty. Urban poverty can lead to not just a crisis of resources but also a threat to social cohesion and societal imbalance.
- By 2050, about 70% of the world's population is expected to live in urban areas. In 2011, 75% of the rural population were concentrated in 19 countries, mostly located in Africa or Asia (with the exception of Brazil, the Russian Federation and the United States). The largest rural population (853 million) is in India, followed by China (666 million).

---

1 The following reports are used to compile the information set presented here: http://www.un.org/ga/Istanbul+5/bg10.htm, http://www.un.org/en/ecosoc/integration/pdf/fact_sheet.pdf and http://www.un.org/en/development/desa/population/publications/pdf/urbanization/the_worlds_cities_in_2016_data_booklet.pdf

62 *Location choice in urban areas*

- Over 60% of the land projected to become urban by 2030 is yet to be built.
- Half of the population of Asia is expected to live in urban areas by 2020, while Africa is likely to reach a 50% urbanisation rate only in 2035.
- Over half of the urban population lives and will continue to live in small urban centres with fewer than half a million inhabitants. While the number of slum dwellers has increased, the proportion of the urban population living in slums in the developing world has declined from 46% in the year 1990 to an estimated 32% in 2010. The world's slum population is projected to reach 889 million by 2020. It is estimated that between the year 2000 and 2010, a total of 227 million people in the developing world were moved out of slum conditions.
- In 2011, Northern America had the highest level of urbanisation (82.2%), followed by Latin America and the Caribbean (79.1%). The rate of urbanisation is expected to be the highest in Africa and Asia over the coming decades. Over the next four decades, the urban population is likely to treble in Africa and to increase by 1.7 times in Asia.

Total population growth

- In 2000, the world population reached 6.1 billion, and is growing at an annual rate of 1.2%, or 77 million people per year. This growth is creating huge pressures on resource uses. The growth is not uniform across the countries. Several countries have grown and are growing at a much faster rate compared to other countries including the ones which have experienced population loss.
- In 1950, 68% of the world's population was in developing countries, with 8% in the Less Developed Countries (LDCs). By 2030, it is expected that 85% of the world's population will be in developing countries, with 15% in LDCs. This shows that the population in large developing countries are expected to grow at a much faster rate. However, the percentage of the world's population that lives in developed countries is declining, from 32% in 1950 to an expected 15% in 2030.
- Much changes are happening within the population structure. Due to medical innovation and a wider reach of modern health care, population is ageing. By 2050, the number of persons older than 60 years will more than triple, from 606 million today to nearly 2 billion. The number of persons over 80 years of age will increase even more, from 69 million in 2000 to 379 million in 2050, more than a five-fold increase.

Urbanisation: Regional comparisons

- The population in urban areas in less developed countries will grow from 1.9 billion in 2000 to 3.9 billion in 2030. But in developed countries, the urban population is expected to increase very slowly, from 0.9 billion in 2000 to 1 billion in 2030.
- The overall growth rate for the world for that period is 1%, while the growth rate for urban areas is nearly double, or 1.8%. At that rate, the world's urban population will double in 38 years. Growth will be even more rapid in the urban areas of less developed regions, averaging 2.3% per year, with a doubling time of 30 years.
- The urbanisation process in developed countries has stabilised with about 75% of the population living in urban areas. By 2030, 84% of the population in developed countries will be living in urban areas.
- Latin America and the Caribbean were 50% urbanised by 1960 but are now in the region of 75%. Though Africa is predominantly rural, with only 37.3% living in urban areas in 1999, with a growth rate of 4.87%, Africa is the continent with the fastest rate of

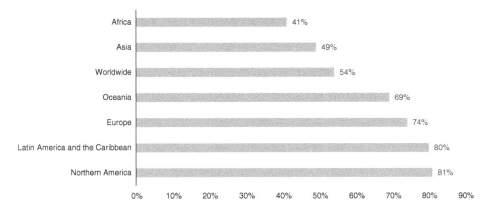

Figure 5.1 Degree of Urbanisation.

urbanisation. In 1999, 36.2% of the Asian population was urbanised and the urban growth rate is in the region of 3.77%. By 2030, Asia and Africa will both have higher numbers of urban dwellers than any other major area of the world.
- Figure 5.1 shows the share of urban population in the total population (2017) for all major regions of the world.

Urban agglomerations, or more megacities

- In 1950, there was only one city with a population of over 10 million inhabitants: New York City. In 1970, Tokyo and New York were the only megacities. Today, there are 13 megacities in Asia, 4 in Latin America, and two each in Africa, Europe and Northern America. Megacities account for a small though increasing proportion of the world urban population: 9.9% in 2011 and 13.6% in 2025. Cities contribute up to 70% of the total greenhouse gas emissions. Urban-based economic activities account for up to 55% of gross national product (GNP) in low-income countries, 73% in middle-income countries and 85% in high-income countries. Cities also generate a disproportionate amount of revenue for governments.
- Of the world's 31 megacities (that is, cities with 10 million inhabitants or more) in 2016, 24 are located in the less developed regions or the 'global South'. China alone was home to six megacities in 2016, while India had five.
- The 10 cities that are projected to become megacities between 2016 and 2030 are all located in developing countries. They include:
  - Lahore, Pakistan
  - Hyderabad, India
  - Bogotá, Colombia
  - Johannesburg, South Africa
  - Bangkok, Thailand
  - Dar es Salaam, Tanzania
  - Ahmedabad, India
  - Luanda, Angola
  - Ho Chi Minh City, Viet Nam
  - Chengdu, China

Table 5.1 Dwelling stock in urban and rural areas

	% Dwellings in urban areas	% Dwellings in rural areas
Australia	88.10	11.89
Chile	86.82	13.18
Canada	79.13	20.54
United States	78.98	21.02
Czech Republic	72.87	27.13
Hungary	72.33	27.67
Finland	71.05	28.95
Lithuania	69.28	30.72
Germany	68.57	31.43
Poland	67.40	32.30
Portugal	65.15	34.85
Mexico	62.98	37.02
Ireland	60.86	38.44
Sweden	59.36	40.64
Japan	57.75	42.25

(Number of dwellings out of the total dwelling stock, located in urban and rural areas respectively, 2015 or latest year available)
Source: Based on data from OECD Questionnaire on Affordable and Social Housing; for a full analysis, refer to the source.

Table 5.1 shows dwelling stock share in urban and rural areas across a selection of major economies.

- In 2000 there were 22 cities with a population of between 5 and 10 million; there were 402 cities with a population of 1 to 5 million; and 433 cities in the 0.5 to 1 million category.
- Some 82% of cities – home to 1.9 billion people in 2014 – were located in areas that faced high risk of mortality associated with natural disasters. Similarly, 89% of cities – home to 2.1 billion people in 2014 – were located in areas that were highly vulnerable to economic losses associated with at least one of the six types of natural disasters.
- On average, cities in the less developed regions were at higher risk of exposure to natural disasters and were more vulnerable to disaster-related economic losses and mortality than those in the more developed regions. Moreover, larger cities tended to be at higher risk of exposure to disasters and more vulnerable to disaster-related economic losses and mortality compared to smaller cities.
- Floods were the most common type of natural disaster affecting cities, followed by droughts and cyclones. These three types of disasters were also the most devastating for city dwellers globally in terms of the mortality and economic losses they caused.

Additional facts and figures about conditions in human settlements from the United Nations Centre for Human Settlements (UNCHS) (Habitat)'s the State of the World's Cities: 2001

Shelter (status as in 2001)

- Seventy-five per cent of the world's countries have constitutions or national laws that promote the full and progressive realisation of the right to adequate housing and 61% of countries in the world have constitutions or national laws that protect against forced evictions.

- Households in cities of developing countries need an average of 8 times their annual income to buy a house; in Africa, they need an average of 12.5 times their annual income, while in Latin America they only need 5.4 times their annual income. The highest rents are in the Arab States, where a household spends an average of 45% of its monthly income on rent.
- One out of every four countries in the developing world have constitutions or national laws which prevent women from owning land and/or taking mortgages in their own names. Customary or legal constraints to women owning land or property are highest in Africa, the Arab States, Asia and Latin America.
- Less than 20% of households in Africa are connected to piped water and only 40% have access to water within 200 metres of their house.

Society (status as in 2001)

- In cities of the developing world, 5.8% of children die before reaching the age of 5 years and 29% of cities in the developing world have areas considered as inaccessible or dangerous to the police. In Latin America and the Caribbean, this figure is 48%.
- In cities of the developing world, one out of every four households lives in poverty. Forty per cent of African urban households and 25% of Latin American urban households are living below the locally defined poverty lines.

Environment (status as in 2001)

- City dwellers in Africa only use 50 litres of water per person per day. In highly industrialised countries, almost 100% of households are connected to piped water. The average water consumption for these households is 215 litres per person daily.
- Less than 35% of cities in the developing world have their wastewater treated.
- In countries with economies in transition, 75% of solid wastes are disposed of in open dumps.
- Seventy-one per cent of the world's cities have building codes with anti-cyclone and anti-seismic building standards based on hazard and vulnerability assessment.
- Buses and minibuses are the most common mode of transport in the world's cities. Cars are the second most common mode used, while walking is the third most common mode. Travel time in Asian cities appears to be the longest with an average of 42 minutes per trip.

Governance (status as in 2001)

- Forty-nine per cent of the world's cities have established urban environmental plans.
- The absolute quantity of local government income varies enormously, with total local government revenue per person in cities of highly industrialised countries being 9 times that of cities in the developing world, 39 times that of African cities and 18 times that of Latin American cities.
- Sixty per cent of the world's cities involve civil society in a formal participatory process prior to the implementation of major public projects.
- Seventy per cent of cities in the world undertake regular independent auditing of municipal accounts.
- Seventy-eight percent of the world's cities publicly announce contracts and tenders for municipal services and 55% of cities have laws that govern disclosure of potential conflict of interest.

The above facts and figures show the huge scale of urbanisation with much complexities in terms of issues and challenges. A concerted policy intervention strategy over a long term along with geo-political stability can bring in changes in these figures for a wider and broader level of prosperity across the regions. With these facts in perspectives, let's look at how cities form and the dynamics of location choice.

Location choice

Choice of location is a key decision made by firms and households. From the firms' perspective, a location should serve well in terms of producing the goods and access to the customers for efficient and timely delivery. This problem can be of a varying nature depending on the industry and type of goods that the firm is engaged with. A firm producing a physical good faces multiple location-specific choices – whether to produce the intermediate products in the location? Whether to produce the intermediate products away at the supply source? Whether to produce intermediate products somewhere between the supply source and final product location? Another dilemma that a firm may face is how to choose a location depending on the customer's location. Think from a supermarket or a grocery store's perspective. A grocery store owner faces competition from other operators. For a customer of the grocery store, a key consideration, besides the quality of products, is how close it is located. This implies that a grocery store owner must consider where the concentration of the customer lies in an urban area. There are traditional factors. However, with globalisation and more integrated economic systems, several other factors have emerged in terms of location choice. In the above discussion, key factors for location choice from a firm's perspective are:

1 **Transport cost**

    This is the most important cost item for a manufacturing firm. Transport of raw and intermediate material and transport of final goods to the customers are key considerations.

    Infrastructure links can influence the transport costs. Both point (e.g. ports) and network infrastructures (roads, railways, bridges) are important considerations for transportation efficiency.

2 **Labour supply and costs**

    Firms need two main factors of production – labour and capital. Labour is a key driver for location choice. Easy access to the required labour pool can make the operation smooth. Quality of the labour pool is important for sustaining the business. More importantly, wage cost plays a significant role in evaluating operational viability in a particular location. If there is a skill gap in a particular location, then firms need to recruit workers from other places and pay for relocation and possibly higher wages, which can be costly. Moreover, willingness of skill-matched workers to move to a firm's location depends on the quality of location from the workers' perspective. The location problem becomes more complicated when location choice variables of the workers are considered along with those of the firm. A worker will evaluate a location's desirability in terms of quality of life, which can comprise several factors such as housing availability and affordability, school quality, recreation opportunities, crime rate and many other factors that are important for raising a family. Moreover, different worker groups may face different choice sets. A firm needs to consider all these in terms of assessing a location's labour market advantage.

    Labour supply and costs change over time and across space. In several emerging countries, labour costs have significantly changed in recent years. Figure 5.2 shows how nominal hourly labour costs vary across Europe and have changed over the recent year.

Location choice in urban areas 67

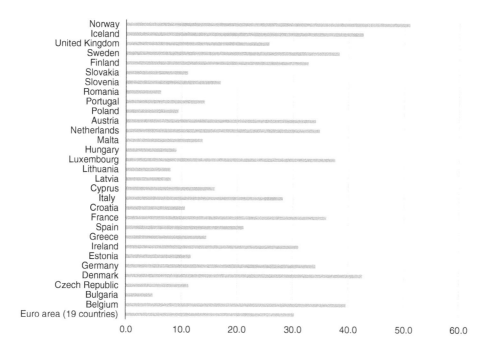

Figure 5.2 Estimated hourly labour costs for the whole economy in Euros, 2016 Enterprises with 10 or more employees.

3 **Land rent**

Cost of the facility including land and building is a significant source of total cost of operation. Land cost differs significantly across locations. Availability of reasonably priced land and built facilities is especially important for firms requiring a large footprint, such as manufacturing units and large service companies requiring a huge office space. Moreover, environmental considerations need to be taken into account while choosing site location. Figure 5.3 shows how premium office rent differed across selected global locations in 2017.

4 **Capital cost**

The ability to raise capital differs significantly across firms. Large firms can tap into national and international capital markets more easily compared to small firms. Regional or local availability of capital is therefore key for small firms. There is significant regional and local variation in credit availability. Figure 5.4 shows the ease of obtaining credit and the strength of legal framework in protecting rights and depth of credit information across world regions.

Energy cost

Energy costs are particularly important. The cost of traditional energy sources have gone up and environmental concerns are driving regulation towards requiring firms to use renewable energy sources. Energy-intensive industries such as steel, metal and petroleum refining require a huge energy budget to operate. Over the last 40 years, energy costs have been very volatile. Increasing concerns of climate change due to fossil fuel consumption and availability of clean energy have led to a shift in the industry practices with technological innovations. Figure 5.5 shows the industrial electricity prices across a selection of major economies in 2016. Figure 5.6 shows the same for industrial gas prices in 2016.

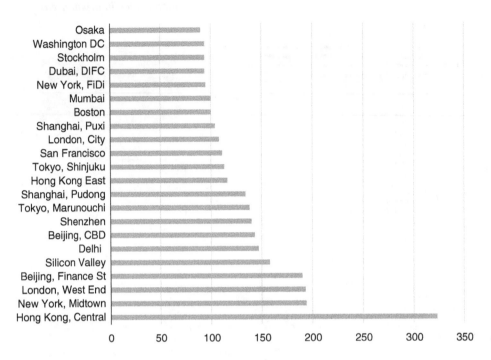

Figure 5.3 Premium office rent (net effective rent in $).

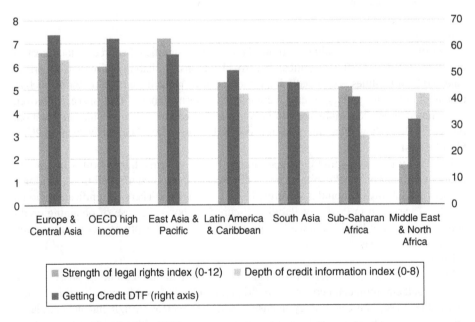

Figure 5.4 Credit availability, 2017.

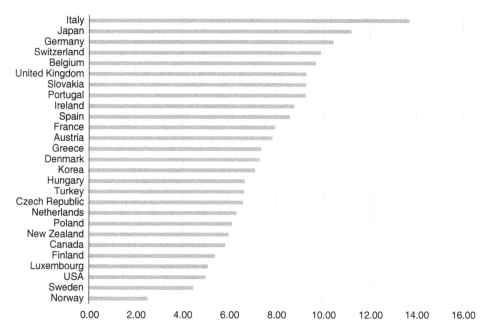

Figure 5.5 Industrial electricity prices in the IEA International Energy Agency (Pence per kWh. 2016).

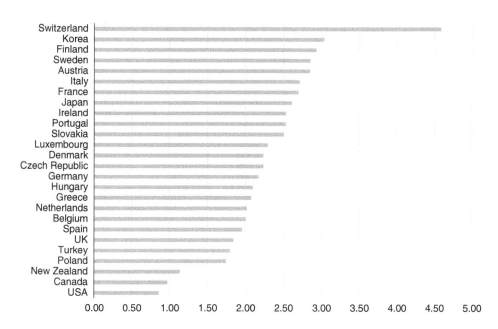

Figure 5.6 Industrial gas prices in the IEA International Energy Agency (Pence per kWh. 2016).

5 **Supply of intermediate inputs**

Most firms depend on inputs that are of an intermediate nature from other providers. Examples include raw materials, parts (e.g. car manufacturer) and a range of business services (e.g. legal, accounting, computing services). Localised availability of intermediate inputs can be a key factor for firms choosing a specific location. This relates to the costs of transport if it is not locally available. Large firms may be somewhat insulated from this factor due to their ability to scale up and build their own intermediate supply chain.

6 **Level of taxes and public services**

A firms' location choice also depends on the extent of taxation at the local, regional and national levels. Taxes have been important considerations for location preference internationally. Favourable tax regimes have been a major driver for business growth in several international locations. This gets intensified if there is also fiscal competition among urban areas, which often leads to incentive schemes in the form of tax breaks, subsidies, low rates and benefits to attract new businesses as well as retain existing businesses. Along with taxes, quality and level of public services become significant factors. The quality of infrastructure and other public amenities are important for the firms and their workers. Table 5.2 shows significant variations in marginal corporate tax rates in the highest rate locations.

7 **Access to knowledge**

Since World War II, several locations around the globe have experienced a concentration of single industry. This is due to the possibility and benefit of knowledge sharing among competitors within a close proximity. If all the major players of an industry are

Table 5.2 Highest marginal corporate tax rates in the world

Country	Top rate	Region
United Arab Emirates	55.0%	Asia
Puerto Rico	39.0%	North America
United States	38.9%	North America
Argentina	35.0%	South America
Chad	35.0%	Africa
Equatorial Guinea	35.0%	Africa
Guinea	35.0%	Africa
Malta	35.0%	Europe
Virgin Islands, US	35.0%	North America
Zambia	35.0%	Africa
India	34.6%	Asia
France	34.4%	Europe
Brazil	34.0%	South America
Venezuela	34.0%	South America
Belgium	34.0%	Europe
Monaco	33.3%	Europe
Saint Lucia	33.3%	North America
Cameroon	33.0%	Africa
Worldwide Average	22.5%	N/A
Worldwide Weighted Average (by GDP)	29.5%	N/A

Source: Based on data from the Tax Foundation; for a full analysis, refer to the source. https://taxfoundation.org/corporate-income-tax-rates-around-world-2016/

Location choice in urban areas 71

in one city region, then knowledge sharing and spillover can benefit all. This is also beneficial for the suppliers and investors in that industry.

Over the last three decades, knowledge sharing and knowledge-based economic development have become key components of success. Countries differ markedly in how they are conducive to formation of knowledge economy. Table 5.3 shows the **Knowledge Economy Index** (KEI) of several countries. It takes into account whether the environment is suitable for knowledge as the key catalyst for economic development. It is an aggregate index – calculated based on the average of the normalised performance scores of a country or region on all 4 pillars related to the knowledge economy – economic incentive and institutional regime, education and human resources, the innovation system and information and communications technology (ICT).

8 **Digital connectivity**

Another aspect that has influenced competitiveness and attractiveness of cities across the world is the internet connectivity. Since widespread use and penetration of the internet, digital connectivity has changed the competitive landscape for the cities. Knowledge-based firms put huge importance on how digitally connected a city or urban area is. A firm that produces a physical good will have a very different set of conditions to be met by the location, compared with a firm that produces a service. While the US leads in this respect, there are several countries which have made significant progress in this regard. All countries are embarking on digital transformation with several economic transactions being performed over digital platforms. This supports innovation, which is key for economic development. Especially in the face of rapid urbanisation and increasing pressure on resources, digitalisation provides a significant efficiency gain from resource utilisation. Table 5.4 shows a significant variation in digital connectivity across the countries. Developed by Huawei, this shows the 2017

Table 5.3 Knowledge Economy Index (2012)

Sweden	9.43	Hungary	8.02
Finland	9.33	Slovenia	8.01
Denmark	9.16	Korea	7.97
Netherlands	9.11	Italy	7.89
Norway	9.11	Malta	7.88
New Zealand	8.97	Lithuania	7.80
Canada	8.92	Slovak Republic	7.64
Germany	8.90	Portugal	7.61
Australia	8.88	Cyprus	7.56
Switzerland	8.87	Greece	7.51
Ireland	8.86	Latvia	7.41
United States	8.77	Poland	7.41
United Kingdom	8.76	Croatia	7.29
Belgium	8.71	Romania	6.82
Iceland	8.62	Bulgaria	6.80
Austria	8.61	Serbia	6.02
Estonia	8.40	Russian Federation	5.78
Luxembourg	8.37	Brazil	5.58
Spain	8.35	Turkey	5.16
Japan	8.28	Mexico	5.07
France	8.21	China	4.37
Czech Republic	8.14	India	3.06

Source: Based on data from the World Bank, Knowledge for Development; for a full analysis, refer to the source.

Table 5.4 Global Connectivity Index

United States	78	Russia	46
Singapore	75	Poland	45
Sweden	73	South Africa	42
Switzerland	71	Mexico	42
United Kingdom	70	Uruguay	41
Denmark	68	Thailand	40
Netherlands	67	Turkey	39
Japan	65	Serbia	39
South Korea	64	Argentina	38
Australia	64	Philippines	35
Germany	63	Egypt	34
New Zealand	62	Venezuela	33
Canada	62	India	33
Belgium	61	Indonesia	33
France	61	Ecuador	31
Spain	55	Kenya	29
Portugal	52	Ghana	29
China	51	Nigeria	29
Czech Republic	50	Tanzania	25
Chile	48	Uganda	25
Greece	46	Pakistan	25
Croatia	46	Ethiopia	23

Source: Based on data from Huawei; for a full analysis, refer to the source. – http://www.huawei.com/minisite/gci/en/

Global Connectivity Index (GCI). The GCI 2017 report shows "three emerging trends from three years of research[2]:

a   Expanding broadband network remains a priority for all clusters as it plays an important role in economic growth. With such capability, nations can identify niche market opportunities based on their unique capabilities and ensure continuous development.
b   Cloud capabilities can act as a powerful equalizer for Adopters and Starters, in particular, to take giant steps ahead in the technology stack, drive innovation and achieve sustainable growth.
c   Investment in ICT Infrastructure initiates a chain reaction of Digital Transformation, with Cloud as a potent catalyst in the chain and a gateway to the power of Big Data and IoT." IoT stands for the Internet of Things.

Models of urban areas

Several simple models of urban areas have been developed. Several key questions are put through these models: where do people choose to locate and live? Where will the firms locate in urban areas? How and why do prices vary spatially within an urban area? What are the spatial and temporal linkages across different locations within an urban area? Answers to these questions are key to understanding the evolution of an urban area or city. It can provide a framework for policymaking to support further economic development.

2   see http://www.huawei.com/minisite/gci/en/

A key ingredient for producing any goods and services is land. Price and rent for land are very important cost items and the decision-making factors for the firms and households. It crucially determines the price and rent of all other built structures. The availability of land can determine what is produced and how much is produced.

In developing a theory for urban land, economists have to simplify the reality to be able to observe changes in determinants as real inputs are brought into the model. The monocentric city model is a standard urban spatial model. It is built on the premise that urban areas evolve around a central place or city centre, i.e. central business district (CBD). Observing patterns of urban development over the history of human civilisation, it is easily recognised that all cities had a central place of attraction with concentrated economic activities and interests. All jobs are located in CBDs. This means people living in the outskirts will have to commute to the CBD every day. As a result of this commute, the whole spatial arrangement of the city is influenced by the CBD location and distances to it. This provides the backbone for the monocentric city model. While cities have become bigger with multiple job locations and the commute has become easier, the monocentric city model is still able to provide answers to several fundamental questions related to where people locate to work and where they choose to live. While several global cities have extended much beyond a single centre, all of those originated around a single place of attraction and concentrated economic interests and activities.

The monocentric model is built on the basic fact that households and firms will choose locations where they will have to commute the least compared to other locations to perform their economic activities. This implies that it is not the absolute location that is important. Rather it is the relative location, relative to other related activities, which is important. Therefore, distant relative locations would be less desirable and thus firms and households would be less willing to pay (e.g. price or rent) for those locations. Economists, therefore, focused on modelling land rent to examine the location choice and evolution of urban areas.

David Ricardo (1821) is the first economist to develop a theory of the land rent. Although he focused on agricultural land, much of the understanding can be transferable to non-farm sectors. The fundamental aspect of Ricardo's model is that agricultural land is not of the same quality in terms of crop productivity and fertility across all available locations. This directly links to the fact of higher desirability and willingness to pay. The urban land market may not be based on fertility of the land but its potential use of the land. Von Thunen (1826) challenged this idea by noting that a very fertile agricultural land at a faraway location may not be that desirable since the cost of transporting the crop to the market can take away all the productivity benefits. This lends direct support to the postulate that land rent will be higher close to the final destination of the product (i.e. crop). The Von Thunen model is directly transferable to urban areas in terms of job locations and choice of residential locations. Key to both Ricardo and Von Thunen models is the scarcity and fixed supply of land of competitive quality. This assumption of fixed supply can also be challenged as more and more land has become available for urban economic activities. Use of land can also be changed, e.g. office to residential, residential to retail.

The fundamental concept to the theory of urban land is bid rent. The housing sector uses a huge portion of that urban land. Bid rent for a household is the maximum rent per unit that the household is willing to pay to occupy that parcel of land. The simplest theories assume that households would travel to work locations in the CBD. So the land closer to the job location will fetch higher rent. The basic location theory based on CBD assumptions predicts that a rent gradient will form with the highest rent per unit in the centre and the lowest on the edge of the urban area. This implies that higher price and rent will attract more development and higher density of living in the centre compared to any other location.

74 Location choice in urban areas

Let's examine how this works with an example[3]:

Greater distance means that the household has to spend more (time and money) to commute to work. So a household would bid less for the parcels of land that are at a greater distance to maintain the utility level. So, bid rent would adjust as the household moves one additional mile.

Marginal Benefit (MB) of moving one mile = (change in bid rent) × (amount of land), which is equal to the Marginal Cost (MC).

Therefore, we can characterise the pricing condition as (McDonald and McMillen, 2011: pp. 89–91):

$$t = -(\Delta R / \Delta x)L \tag{5.1}$$

Where t is the commuting cost (time + money); ΔR is the change in bid rent R as distance x changes by ΔX, *which would be a negative* number given the intuition.

$$(\Delta R / \Delta x) = -t / L \tag{5.2}$$

Equation (4.2) is the slope of the bid-rent function, which is equal to -1 times the MC of distance divided by the amount of land the household would occupy.

Let's take a quantitative example (following McDonald and McMillen, 2011):

Suppose the annual cost of commuting an extra mile = £100
A total of 1,000 miles will need to be added to the annual commuting.
The household occupies 1,000 square feet of land.
So, the slope of the bid rent function = £100/1,000 = 10 pence per sq. ft per mile;
For an acre of land, i.e. 43,560 sq. ft, it is £4,356 per acre per mile.

However, significant complications would arise when the real possibility of different households bidding differently for the same parcel of land is added. This is due to the fact that different economic interests carry different value to different households. Such heterogeneity can add a huge amount of complication to the bid rent theory. Figure 5.7 shows how three households can differ in their bid rent functions (BR1, BR2, BR3), although all indicate a negative relationship between the bid rent and the distance to be commuted.

It is now possible to derive the shape of the bid-rent function (refer to a more complete derivation in McDonald and McMillen, 2011, pp. 94–95):

Following equation (5.2), take the first derivative with respect to distance (x).

$$\frac{dR}{dx} = \frac{-t(y)}{L(y)} = -t(y).L(y)^{-1} = R' \tag{5.3}$$

Next, take the first derivative of equation (5.3) with respect to income (y).

$$\frac{dR'}{dy} = -t'(y).L(y)^{-1} + t(y).[L(y)]^{-2} L'(y) = -\left(\frac{1}{L(y)}\right)\left[t'(y) - \frac{L'(y)t}{L(y)}\right] \tag{5.4}$$

---

3 McDonald, J.F., McMillen, D.P. (2011). *Urban economics and real estate: Theory and policy*. John Wiley & Sons.

Location choice in urban areas 75

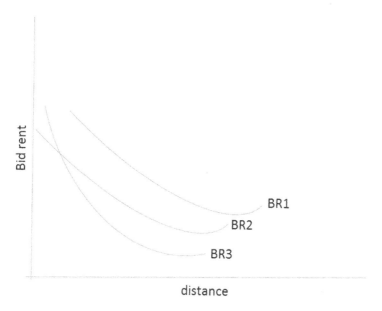

Figure 5.7 Bid rent curves.
Source: Adapted from McDonald and McMillen, 2011.

Rearranging and replacing in equation (5.4), we get:

$$\frac{dR'}{dy} = \frac{-t}{yL}\left[\frac{dt/t}{dy/y} - \frac{dL/L}{dy/y}\right] = \frac{-t}{yL}[E_{ty} - E_{Ly}] \quad (5.5)$$

So, the shape of the bid-rent function in equation (5.6) relates to the income elasticities of commuting cost (E_{ty}) and demand for land (E_{Ly}).

The shape of the bid-rent function can tell lots of stories! For the residential sector:

- Bid-rent reflects tastes and preferences of households. It can thus shape communities as 'like-minded' households with similar tastes and preferences may choose similar locations.
- Bid-rent function considers positive and negative feedbacks to the household choices.
- The factors that positively influences the residential bid-rent are, e.g.: access to downtown employment, shopping and other urban amenities like schools, health services, fire and police services.
- The negative factors include, e.g. crime, environmental pollution, congestion, noise pollution and taxes. Note the tax can have positive effects as tax revenue is spent on providing and funding public services.
- Different types of households exhibit different bid-rent functions, e.g. single adults or couples without children may make relatively higher bids for land located near the downtown area but may make lower bids for suburban areas. However, the families with children will bid relatively higher for spacious and safe suburban locations.

76 *Location choice in urban areas*

- Similarly, businesses can differ significantly in their bid rents. Some businesses, such as service firms like financial institutions, law firms etc are willing to pay much higher for locations in the downtown or high street area compared to other firms. Some firms who require more direct contact with the customers and other economic agents would bid higher for the central locations. Some manufacturing firms who do not require daily contacts and interactions with customers may not bid highly for the CBD locations, e.g. heavy industries like steel plants and fertiliser manufacturers.

So the question is: how do the planners reconcile all these heterogeneous bid rent functions so that they are able to take decisions on public services and land uses in the area? Competition and market forces can determine the solutions but some interventions are needed for meeting the objectives of the city authorities and preserving aspects of broader societal goals.

Land-use planning

Land-use planning can be defined as regulatory interventions by the state into development by controlling land-use change through infrastructure planning (such as transport links, water supply, etc.). This often involves specifying which developments are permissible at specific locations leading to active 'spatial planning'. While there are often sound motivations and reasons, they can have negative implications on regulation that can hinder development or channel it in particular directions. These regulations are separate from building regulations which concern the safety and performance of built structures. Typically, land use planning are governed by local governments and sometimes through special purpose agencies. Locational distribution in terms of governance are determined by administrative boundaries. Due to much granularity in administrative territories, in some countries like the UK, it can lead to many authorities, e.g. there are 418 principal (unitary, upper and second tier) councils in the UK[4] – 27 county councils, 201 district councils and 125 unitary councils. In some areas of England, local government is divided between a county council (upper tier) and a district council (lower tier), which are responsible for different services. In other areas, there is a single unitary authority instead.

In Scotland, Wales and Northern Ireland there are only unitary, single tier councils. As households often cross administrative boundaries to fulfil their economic interests, some spatial spillovers (or leakages) can crop up which may also lead to policy conflicts among the authorities and undesirable delays in realisation of economic and social benefits. Moreover, often a hierarchy of national, regional and local planning authorities operate to govern land-use planning issues. In Europe, local/regional structures are common in Europe. In the US, federal, state and local governments work towards developmental goals. There are also public agencies in charge of certain aspects of public services which require co-operation among local authorities and agencies, e.g. transport and environmental agencies.

Administration of land-use planning involves multi-layer decision-making and it is accomplished through several ways. Zoning is a key approach which sets out zones for land-use types (e.g. residential, commercial, mixed use) and often requires rule-based, prescriptive framework, e.g. development of maximum height, maximum allowable density, impact mitigation tools, land value capture tools, etc. Detailed evaluation of a competing proposal is undertaken to permit or reject development proposals. Alignment with the master plan is

---

4 https://www.lgiu.org.uk/local-government-facts-and-figures/

required for the big projects with the small project being decided on a case-by-case basis. Countries differ in terms of custody of development rights. Central and state governments often carry responsibility of overseeing this. Private monopolists can also get such privileges.

Economics of land-use planning

There has been much work on using various economic theories and tools to analyse planning issues, impacts of policies, evaluation of policy alternatives, etc. This links to the seventh unique attribute of 'Political Economy' that we discussed in Chapter 2. There has also been much debate around positive and negative effects of land-use planning. Both positive and normative economics have been applied to these issues. Much of the motivation has been drawn from the objectives of optimisation of social welfare with marginal social benefit (MSB) being equal to marginal social cost (MSC). There has been a growing recognition of the impacts of planning systems in countries with a strong and long history of land-use planning such as in the UK. Some of the arguments has centred on how the breadth and depth of planning systems can stymie the ability to use the housing supply to counter house price inflation and affordability concerns. It can also restrict availability of land for new housing supply. Both in the US and UK, there have been much discussions on making processes simple and efficient as it can otherwise delay the development process and add significant uncertainty in the development through long lags and uncertainty crippling decision-making ability by the developers. In the UK, the Barker Review was conducted to evaluate land-use planning in 2006[5]. As Barker commented in the report:

> Planning policies and decisions require the consideration and weighing of a range of factors: local, regional and national interests, environmental issues and economic growth. In part, planning policy is a set of regulations aimed at correcting market failure – in which context climate change is becoming increasingly important. But planning should also play a central role in delivering the vision that regional and local government has for its area, and it should enable development to fulfil that vision. This also implies that the planning system needs to be accessible to the community, and that community engagement should take place at the right time, when development plan documents are being drawn up and before major new development takes place.

One of the main recommendations from the review was to make the planning system more sensitive to market dynamics and promote more efficient use of land.

The biggest challenge in applying positive economics to land-use planning is the fact that we are often dealing with public goods with complex externalities and sometimes with a high degree of publicness, i.e. non-excludability in consumption. As Barker aptly put it, "it is inevitable that some individuals and groups will be adversely affected by particular planning decisions, and in that sense planning will always be controversial." Therefore, two major objectives are – controlling land-use rights and scale as well as location of infrastructure. Being public in nature, the non-excludability feature give rise to conflicts in land-uses and free-riding behaviour on public services and local amenities. Optimisation of provision of public goods becomes quite challenging when faced with externalities along with the objective of preserving the amenities that cannot be priced in the market, e.g. clean air, natural

---

5 Barker, K. (2006) Barker review of land use planning: Final report, recommendations. The Stationery Office.

environment, cultural history, nature disaster-linked losses, other intangibles, etc. This may also have actions being taken to protect neighbourhoods or building features such as the creation of conservation areas, zoning controls and regeneration of urban areas. Such actions can lead to many implications for the urban form. For example, transportation links and related infrastructure can lead to commuting behaviour change, clustering of population and social infrastructures (such as school and hospitals), which in turn can shape much of the city expansion effects and geographic size. It is often argued that the development companies should bear the cost of such network effects through charging developers and users with land-value capture tools such as taxes, impact fees, specific provisions for infrastructure built, owning and financing of a part or all of infrastructure intervention. Many cities around the world have adopted various forms of land-value capture.

Land-use planning can often be analysed through various approaches recommended by economic theories such as:

- Positive welfare economics:
  Under this approach, government's role is of a social welfare maximiser with clear goals for protecting public goods and controlling or shaping consumption behaviours. A key target is to minimise the externalities through democratic processes with wide buy-ins from the residents. However, beneficiaries may not always be required to compensate the losers as the voting processes are not perfect under wider social choices.
- Pigouvian tax:
  Pigouvian tax is a tax on market-based activities that may generate negative externalities that is not included in the market price, which leads to inefficient market outcomes. This is done by equating to a level where MSC = MSB. Congestion taxes are good examples of such taxes. Winners compensate losers through government mechanisms, e.g. S106 provisions in UK, impact fees. There are three downsides of these taxes: (a) it requires strict regulatory framework and monitoring, (b) it is difficult to quantity MSC, MSB and subsequently, the optimal tax rate, and (c) if not monitored, it can also lead to sub-optimal changes in behaviour.
- Local political economy:
  In a democratic set-up, local governments consist of representatives elected through democratic processes. The voting process is undertaken by both utility-maximising groups of voters and elected officials. Most land-use policies are decided and actioned by a voting system. Often, policy issues and legislations are brought forward by voters' preferences. Any dominance of specific interest groups can often lead to outcomes that may not reflect the wider population's interest. Such misalignment of interests becomes complex with politicians maximising their self-interests of staying in power. As a result, populist measures can sometimes dominate over the welfare maximising policies. In the UK, NIMBYism (not-in-my-backyard) is often referred to as a major deterrent for flexibility in the housing supply. The voting power and lobbying prowess of homeowners can significantly shape the politician's and elected official's agenda and actions which may compromise on achieving social welfare goals. While property owners stand to gain from property price inflation due to rising demand and heavily constrained supply, worsening affordability can lead to social welfare loss. Beneficiaries and losers may be spatially arranged in different ways.
- Tiebout model:
  Tiebout (1956) analyses local areas/municipalities providing a basket of goods (government/public services) at varying prices (tax rates) across the areas. Given that

individuals' valuations of the public services and ability to pay taxes differ, individuals/households will move from one local community to another, until they find the one which maximises the utility. The Tiebout framework predicts that the bidding process will determine an equilibrium provision of local public goods, thereby 'sorting' the population into communities. The Tiebout model is built on a set of basic assumptions, especially on labour market linkages with the housing market, which can significantly undermine the efficacy of the model if the assumption is relaxed. There may also be significant free-riding behaviour, which can exacerbate inequalities.

- Coasian framework:

    Coase (1937) put forward a compelling argument that if we are able to define appropriately property rights, externalities can be sorted with effective bargaining between private economic agents with no interference from public bodies. A key requirement for such solutions to emerge is that transaction costs must be low and not prohibitive at all. An appropriate, robust institutional framework can ensure such solutions to be successful. There are some arguments in the literature that planning provides that framework and acts as the forum for the Coasian bargaining process.

Challenges for land-use planning

Land-use planning processes are fraught with significant challenges across the world. While some challenges are common to all areas, there are local-level factors that can shape land-use planning. Public authorities often grapple with a multitude of the following challenges:

Spatial planning requires deep contextual understanding for decision-making on public amenities and future planning. To be effective and sustainable, a huge amount of information is required. The information would need to be sourced from many sources with much heterogeneity in the types of information. Therefore, ideally, the required information set would be big and complex. In order to minimise the 'stochastic' nature of spatial planning as it is inherently uncertain due to uncertain events and the unpredictable nature of economic cycles, it is often recommended that planning arguments are fact-based, avoiding judgement-based decision-making. More and better data capture and processing can help reduce the 'stochastic' or uncertain nature of spatial planning. However, this is not easy to accomplish. It requires appropriate data processing along with appropriate data protection protocols, in addition to collection and compilation of data.

Spatial externalities, being complex and omnipresent, contribute greatly to this challenge. Numerous conflicting interests among private economic agents make it impossible to reach consensus without state interventions. Political, cultural and socio-economic conditions across different countries lead to different views towards such interventions. Therefore, achieving Pareto solutions are difficult with many winners as well as losers, which can call for 'compensation and betterment', e.g. impact fees, local development incentives, etc. as compensation measures as well as land value capture tools as 'betterment' taxation. 'Betterment' refers to benefits which a land or property owner receives due to public activities on an adjacent expropriated parcel (externality). Several seminal works have contributed refinement of this concept starting with the original idea from George (1879). The Expert Committee on Compensation and Betterment (Uthwatt), in 1942 British Planning system reform, described 'betterment' as any increase in the value of land (including the buildings thereon) arising from central or local government action, whether positive, e.g. by the execution of public works or improvements, or negative, e.g. by the imposition of restrictions on other land. Several attempts have been made to devise appropriate instrument for capturing betterment and structure compensation but with mixed acceptance and reviews. Such challenges get complicated with political influence and voting mechanism of homeowners as explained before.

Housing and neighbourhood

A key consideration of location choice is neighbourhood quality indicators. House prices reflect valuation placed on physical units and the premiums/discounts associated with the locational attributes. With location coming along, the choice set becomes complex with several factors interacting at the same time and in complex combinations. As we have seen before, depending on their tastes and preferences, different households would bid differently for the same house and parcel of land. That means systematic differences in the households' bids would entail 'clustering' of certain types of households, i.e. first stage of forming communities based on multi-dimensional choice set. Then the relevant question becomes: How do households 'sort' among multiple locations in an urban area or get 'sorted' among the local areas?

A well-developed literature exists in this regard. Ross and Yinger (1999) provide a comprehensive review of this literature and summarises the body of works at the intersection between urban economics and local public finance, considering both a housing market and the market for local public services. Three related questions are explored:

1 How does the housing market allocate households to communities when local public services and taxes may vary from one community to the next?
2 How do communities select the level of local public services and tax rates?
3 Under what conditions are solutions to the first two problems compatible, i.e. when does an urban equilibrium exist?

The research works have developed a consensus model of the allocation of households to jurisdictions based on bid functions and household sorting, as well as alternative approaches to this issue along with a critical evaluation of models of local tax and spending decisions and models in which both housing and local fiscal variables are endogenous. It reviews empirical research evidence with a focus on tax and service capitalisation on household heterogeneity within jurisdictions, and on the impact of zoning. They also consider normative theories about a decentralised system of local governments, i.e. to what extent decentralised systems may lead to efficient allocations of households to communities or efficient local public service levels, and it discusses the fairness of local public spending. In general, Ross and Yinger argue that the bidding/sorting framework is strongly supported by the empirical evidence, which corroborates many works starting with the seminal work of Tiebout (1956) framework. Overall, policies suffer from biases driven by misallocation of households to communities (misplaced 'sorting'), the property tax capitalisation, public service capitalisation and heterogeneity in choice sets and outcomes. It is a daunting task for any policy to eliminate these biases. Let's look at the Tiebout (1956) framework and how it provides a useful lens to analyse the above issues.

Tiebout (1956) analyses local areas or municipalities providing a basket of goods (government/public services) at varying prices (property or council tax rates) across the areas. Given that individuals' valuations of the public services and ability to pay taxes differ, individuals or households will move from one local community to another until they find the one which maximises the utility. The Tiebout framework predicts that the bidding process will determine an equilibrium provision of local public goods, thereby 'sorting' the population into communities. However, the Tiebout model is built on a set of basic assumptions, which can be challenged:

- Consumer-voters are fully mobile and will move to that community where their preference patterns, which are set, are best satisfied.
- Consumer-voters are assumed to have full knowledge of differences among revenue and expenditure patterns and to react to these differences.
- There are a large number of communities in which the consumer-voters may choose to live.
- Restrictions due to employment opportunities are not considered. It may be assumed that all persons are living on dividend income.
- The public services supplied exhibit no external economies or diseconomies between communities.
- All households are homeowners.
- Local public services are financed through a local property tax setting assessed value equal to market value.

It provides an excellent framework to start asking several pertinent questions:

1 What is the public choice mechanism? This can be based on the median voter framework. But the question arises when we consider the strong possibility of voters being myopic. Do voters understand that their decisions may affect housing prices? Do voters understand that their decisions may affect housing consumption, both their own and that of other households in the community?
2 Are voters owners or renters? Renters add significant complexities as taxpayers and having voting rights as choice sets and preferences, compared with homeowners, can be significantly different.
3 What is the technology of public production? There are three aspects to consider: returns to service quality, the degree of publicness and the impact of community characteristics.

Neighbourhood choices and location of households leads to *'coalescing'* of similar tastes and preferences, i.e. households are finding locations, knowingly or unknowingly, where 'own' people are driven by a sense of *'identity'*. Such behaviour and socio-economic outcomes over a long period of time can lead to segregation by social, economic, demographic and racial status, caused by tastes and preferences for social linkages. There can very well be some external influences such as regulations, diktats, discriminatory policies, etc. Regardless, segregation can lead to mixed and negative effects. At some level, it can facilitate networking and some benefits of networking but also, considering longer run outcomes, it may lead to sub-optimal solutions in terms of housing and labour market outcomes.

Housing and education

The nexus between housing and educational outcomes (as a locational characteristic) has been studied in great detail by many studies. Across many countries, access to education or local schools is determined by location of the houses. Being a resident provides access to local public schools. If the quality of the local school is good, that adds to the list of housing demand shifters. Typically, the effect of school district performance are measured by student test scores and ranks and the analysis controls for the effect of the student socio-economic and demographic composition on local property values. Overall, studies find a positive effect. But it is not straightforward. Questions crop up regarding: which school attributes? Whether it is the school's performance in tests – like in some public exams – or the 'demography' of the student and their parents that matter. The relationship can also be dynamic – 'chicken-egg' problem – leading to severe endogeneity bias. How do we *'dichotomise'* the effects?

The education systems, especially the public institutions, around the world shows significant heterogeneity in terms of performance, students' socio-economic status, and racial and ethnic compositions. For example, the US and UK show significant variation across school districts (US) or attendance zones (UK).[6] There is a large literature on the issues related to the segregation of students by socio-economics status, race or ethnicity and the extent of equality of educational opportunity. In the US, racial segregation has been shown to have a strong relationship with lower levels of educational outcomes (possibly subsequent economic outcomes) for minority students (see analysis by Hanushek et al., 2003; Mickelson, 2003) and/or lower school quality (Freeman, Scafidi, and Sjoquist, 2005). Such issues also have been shown to have effects on the property values within the neighbourhood (around the school) and also the surrounding neighbourhoods. Housing seems to determine the access to good education. Weimer and Wolkoff (2001) examine the capitalisation of public school performance and tax costs. Their findings confirm the influence of school quality, especially elementary school, on house prices: increases in incomes of student families increase house prices.

However, care must be taken in terms of spurious results as the school quality or performance may be correlated with unobserved neighbourhood characteristics that are not contained in the dataset and those correlations, if strong and significant, can have huge implications and may easily lead to wrong inferences. Researchers have tried to address this concern by innovative techniques. Although questions do remain, the studies have been reasonably successful in deriving fairly confident results with convincing interpretations. For example, Black (1999) addresses this concern by examining a sample of housing transactions that occur on the boundary between elementary school attendance zones within the same school district. Similarly, Gibbons and Machin (2003) and other studies used the same technique to effectively control for the unobserved heterogeneity. However, the boundary technique may not always work if the school district boundaries share the same with the town or they end up dividing heterogeneous neighbourhoods. This implies that houses on either side of a boundary can actually belong to different neighbourhoods, which would not work in terms of effective control for unobserved heterogeneity.

Another issue is that parents may also care about the racial composition of the school. Clotfelter (1975) examines the effect of school desegregation on property values finding that the schools that experience the greatest declines in segregation during the 1960–1970 period also experienced relative declines in house prices. However, Norris (2002) finds no effect or possibly a positive effect of African-American representation on house prices. Weimer and Wolkoff (2001) also obtained a positive effect of poverty on house prices. The results have been one of a very mixed bag. As mentioned, a big challenge in these analyses is effective control for the unobserved heterogeneity. Black (1999) relies on the assumption that the attendance zone boundaries do not change. However, Cheshire and Sheppard (2004) argue that attendance zones may change.

---

6 This review is drawn from Clapp et al. (2008). For a full review, see the paper – Clapp et al. (2008: pp. 452–454) – and other related literature.

Table 5.5 Example of hedonic estimation results: Neighbourhood effects

Math test score	0.074 (8.99)
Fraction of students enrolled in free lunch program	−0.363 (−6.19)
Fraction of students Non-English Speakers	0.419 (4.93)
Fraction of students African-American	0.279 (6.64)
Fraction of students Hispanics	0.128 (1.22)
Effective property tax rate	−0.531 (−11.49)

The dependent variable is the natural log of house price, t-stats are in parentheses.
Model includes several other variables. Refer to the full table.
Source: Adapted from Table 3 from Clapp et al. (2008).

Clapp and Ross (2004) find that the effect of test scores varies depending on school demographic attributes using a panel data analysis. Clapp and Ross (2004) study a simultaneous process by analysing how property values, performance and demographic characteristics of schools evolve over time. But they impose a key identification restriction that changes in the Labour Market Area (LMA) price level affect changes in the individual district price levels, but not changes in district demographic composition, except through the variation of school district housing prices for different market segments.

Clapp et al. (2008) used a sample of sales of owner-occupied properties with one to three units spanning over 11 years from 1994 to 2004 across towns in the state of Connecticut, US. The sample has 356,829 transactions for one to four unit owner-occupied structures, after any filtering to eliminate invalid or non-representative transactions. The model controls for time-invariant neighbourhood (as defined by census tracts in the US) attributes using fixed effects. The key advantage over the earlier boundary analyses – the effect is driven by a complete sample of housing units rather than only those along the boundaries. In general, they find that people in the state of Connecticut during the study period seem to be more concerned about the changes in socio-economic and demographic attributes over time than the changes in test scores over time when deciding to be home-owners. The conundrum continues. However, there is little doubt on the effects education can have on housing outcome or values but the extent and direction vary depending on the context and is highly fraught with estimation challenges within a complex dynamics.

The hedonic framework, as discussed before, is typically used for such analyses mentioned above. Neighbourhood level characteristics are studied within the hedonic specification. Table 5.5 shows an example of a typical specification for neighbourhood attributes from Clapp et al. (2008).

Housing and neighbourhood attributes

Similar to the previous example of analysing the effect of school quality, many other neighbourhood level attributes can be studied in terms of their contribution (premiums or discounts) to housing demand and thus, house price changes. We cannot possibly list all the factors that determine the price. But, let's try to list at least some!

1. Property tax rate: homeowners and renters pay property tax (or council tax in the UK). The property tax revenue is typically used by the local public authorities to finance public education, amenities such as fire and police services, waste collection services, some local infrastructure. While it can lead to negative effects for housing demand as this is deemed to be a 'cost' for the homeowners and renters, it can also be viewed favourably

as the higher level of property tax revenue can help financing of local public amenities and education.
2 Neighbourhood income, size of houses, etc.: per cent of owner-occupied housing, income level, quality and size of houses affect housing demand and prices. Typically the effects are positive. Rating of the quality of houses next door and on the block.
3 Air pollution: quality of air and level of pollution have negative impacts. However, if we bring in the labour market dynamics, the net effect may not be clear, as some polluted areas may also have a higher level of employment.
4 Proximity to highway interchange (positive effect) – this can also have dampening effect on demand due to heavy traffic, noise; similarly, adjacent to rail line, highway, or transit line (negative effect).

The above list is a short list. There can be many other neighbourhood attributes having significant impacts on housing demand. It is important to note that effects are not always clear and it is possible to have direct and indirect effects as well as positive and negative feedback effects. Therefore, all neighbourhood effects should be studied carefully. The presence of significant neighbourhood effects that cannot be observed in the data can lead to severe biases in empirical conclusion, which is a frequently cited problem in the literature. Effective empirical and econometric mechanisms need to be put in place to draw meaningful and robust inferences.

In this chapter, I have discussed the locational choice issues with respect to the housing market. I have highlighted the areas of research in education and other neighbourhood characteristics. All these locational choice issues lead to significant search effort by the households as they choose among alternative locations. In the next chapter, I discuss the housing search issues, tenure choice and aspects of housing mobility.

Selected research topics

- Bid rent and changing economic geography due to digitalisation.
- Ageing population and location choice.
- What would be the likely impact of the following aspects on housing demand and prices?
  - Airport noise; Proximity to contaminated area; Proximity to nuclear power plant; Industrial noise;
  - Crime rate in area;
  - Heavy traffic on street; Location on floodplain; Distance to employment; Distance to shopping; Distance to airport;
  - Proximity to a religious place; Proximity to a (nice) park;
  - Existence of zoning and conservation areas;
  - Previous price increases in neighbourhood to capture future price expectation.

6 Housing tenure, search and mobility

Chapter outline
- User cost of housing
- Tenure choice
- Rent control
- Housing search model and role of intermediaries
- Regulation and policy issues in brokerage industry
- International comparison of brokerage industry
- Online platforms
- Selected research topics

In Chapter 4, we discussed house price estimation in detail. One of the significant challenges in the house price estimation framework is how to incorporate housing costs appropriately. One of the arguments is that fluctuation in house prices is faced by mortgage-paying homeowners and renters (Crawford and Smith, 2002), and therefore, monthly rental payment and mortgage instalment should be incorporated in the house price index. This implies homeowners without mortgage obligations do not face any housing costs! That is not quite the case if we take into account monetary and non-monetary costs and benefits, especially if viewed within the opportunity cost perspective. This is referred to as 'user cost of housing'. There is a well-established literature on user cost of housing.[1] This framework is often utilised to examine tenure choice decisions of households. In general, a household would rent if the user cost of housing is higher than the cost of renting. In this chapter, we first present the user cost of housing framework and discuss tenure choice decisions based on that understanding.

Díaz and Luengo-Prado (2012) show a framework through which we can examine the components and issues related to the user cost of housing. They note that

> from the mid-90s to 2007, we witnessed a substantial rise in real house prices and no significant decreases (and even modest increases) in homeownership rates. For example, in the US, the homeownership rate increased from 66.27 to 68.15% from 1998 to 2007, while houses appreciated around 60% in real terms over the same period. The question that arises is why this is the case when, at first sight, it may appear that owning a home becomes relatively more expensive than renting when house prices go up as rental prices

---

[1] For a review of the literature on this topic, see "User cost, home ownership and house prices: United States," A. Díaz and M.J. Luengo-Prado in The International Encyclopedia of Housing and Home, eds. S.J. Smith, M. Elsinga, L. Fox-O'Mahony, S.E. Ong, and S. Wachter. 2012. Amsterdam: Elsevier.

tend to move only sluggishly with house price increases. Why did homeownership not decrease significantly as houses were becoming more expensive?

House price is not the only costs and there are significant benefits which are not directly observable and may be non-monetary, nevertheless enters the decision-making framework of households choosing to become homeowners or stay put in their rental homes. Homeowners face significant risks – credit risks, transitory income risk, wealth depletion risk, all these call for appropriate compensation for increased risk as a homeowner. Díaz and Luengo-Prado (2012) derive the user cost of housing from a simple model of housing as the shadow price of housing services. However, their user cost definition is different, as acknowledged by the authors, from that used by Poterba (1984), Himmelberg et al. (2005), and Poterba and Sinai (2008). The differences are: (1) transaction costs are considered, (2) they differentiate the cost of own money invested in a house from the cost of borrowed money, i.e. mortgage. This allows user costs to vary across households in the same location depending on the scale of leverage.

In order to derive user costs, we need to treat homeowners as both consumers and landlords. Individuals maximise life-time utility from housing and a composite consumption good subject to budget and wealth constraints. If we consider that households have two items in their consumption basket – housing (H) and a non-housing composite good (C), the lifetime utility is a function of H and C:

$$\max(lifetime\ utility) = f(H,C)$$

$$P_H/P_C = f(rent/P_C, UCC) \qquad (6.1)$$

Formally, the user cost of capital builds the following factors into housing demand:

- Interest rates – higher interest rates lower housing demand and vice-versa.
- Capital gains – house price appreciation create capital gains, wealth effect and vice-versa.
- Future price expectations also matter and is an important consideration for the homeowner, which can drive much of the investment demand of housing.
- In the calculation, we also need to account for inflation, tax and depreciation factor or maintenance costs.
- Higher interest rate and stricter credit conditions raise user cost.
- Higher price expectations reduce user cost.

Himmelberg et al. (2005) show an application of user cost model that can evaluate whether the level of housing prices is 'too high' or 'too low'. They note that the key problem in the traditional measures of overheating in housing markets is the purchase price of a house being viewed as the annual cost of owning. A more complete reflection of the financial return associated with an owner-occupied property should compare and incorporate the 'imputed rent', i.e. the cost to rent an equivalent property with the lost income that one would have received if the owner had invested the capital in an alternative investment – the 'opportunity cost of capital'. This computation should take into account differences in risk, any tax deductibility benefits (although not all countries have similar tax benefits) from owner occupancy, property taxes or council taxes (note that in some countries renters pay council tax), maintenance expenses to preserve the capital value of the property or slow any physical depreciation, and more importantly, any expected capital gains from owning the home, although this could also be a loss. Himmelberg et al. (2005) note that the true one year cost of owning a house (the 'user cost') can then be compared to rental costs or income levels to judge whether the cost of owning is out of line with the cost of renting, or unaffordable at local income levels. They provide a review of the most commonly used procedure for calculating the annual cost of homeownership, its implications for house prices and the assumptions.

The formula for the annual cost of homeownership, also known in the housing literature as the 'imputed rent', is the sum of six components representing both costs and offsetting benefits (Hendershott and Slemrod, 1983; Poterba, 1984; Himmelberg et al., 2005).

$$\text{Annual Cost of Ownership} = P_t r_t^{rf} + P_t \omega_t - P t \tau_t \left(r_t^m + \omega_t \right) + P_t \delta_t - P_t g_{t+1} + P_t \gamma_t \quad (6.2)$$

In equation (6.2), P_t = house price; r_t^{rf} = risk-free interest rate; ω_t = property tax rate (or council tax in England); τ_t = tax deductibility of mortgage interest (not in all countries); r_t^m = mortgage interest rate; δ_t = maintenance cost (fraction of home value); g_{t+1} = expected capital gain (or loss); γ_t = additional risk premium to compensate homeowners for higher risk of owning vs. renting. Annual cost of homeownership = sum of the following cost and benefit components (as indicated in the equation above):

- the foregone interest ($P_t r_t^{rf}$)
- property taxes ($P_t \omega_t$)
- tax deductibility of mortgage interest ($P t \tau_t (r_t^m + \omega_t)$), i.e. effective tax rate on income times the estimated mortgage and property tax payments
- maintenance costs ($P_t \delta_t$)
- expected capital gain (or loss) ($P_t g_{t+1}$)
- risk premium to compensate owners ($P_t \gamma_t$)

In this framework, Himmelberg et al. (2005) argues that "a house price bubble occurs when homeowners have unreasonably high expectations about future capital gains, leading them to perceive their user cost to be lower than it actually is, and thus pay 'too much' to purchase a house today." At the equilibrium, the expected annual cost of owning a house should be equal to the annual cost of renting, i.e. 'tenure neutrality' is achieved. If annual ownership costs outpace the increase in rents, house prices would need to adjust downwards for prospective homeowners to decide to buy instead of renting. They would rather opt for homeownership if annual ownership costs fall. Such an adjustment process can lead to 'no arbitrage' condition where one-year rental costs must equal the annual costs of owning as defined by equation (6.2). In terms of per unit user cost:

$$u_t = r_t^{rf} + \omega_t - \tau_t \left(r_t^m + \omega_t \right) + \delta_t - g_{t+1} + \gamma_t \quad (6.3)$$

i.e. rental cost = R_t = price of house x $u_t = P_t u_t$

i.e. $P_t/R_t = 1/u_t$; which implies at the equilibrium, the price-to-rent ratio should equal the inverse of the user cost. Thus, fluctuations in user costs due to any policy and market interventions (for example, by changes in mortgage interest rates, first-time buyer's tax subsidy, stamp duty waiver, etc.) lead to predictable changes in the price-to-rent ratio that may reflect fundamentals, not bubbles, which are more to do with misplaced positive expectations regarding future capital gains.

88 Housing tenure, search and mobility

Similarly, there is a large literature on the determination of UK national house price equations, e.g. Muellbauer and Murphy (1997), Meen (2002), based on the standard life-cycle housing model. Since the model is well known, only the key equations are presented here. The dynamic life-cycle model yields the user cost of capital. This allows for the longevity of housing and, so, it can be used as an asset. Therefore, it takes account of the expected capital gain on housing, the marginal rate of substitution (MRS) between housing (H) and a non-housing composite good (C):

$$\mu_H / \mu_C = g(t)\left[(1-\theta)i(t) - \pi + \delta - \dot{g}^e / g(t)\right] \qquad (6.4)$$

where: μ_H / μ_C = user cost of capital
$g(t)$ = real purchase price of dwellings = P_H / P_C
θ = household marginal tax rate
$i(t)$ = market interest rate
δ = depreciation rate on housing
π = general inflation rate
(.) = time derivative
$\pi + \dot{g}^e / g(t)$ = expected nominal capital gain on housing;
δ, θ are assumed to be constant;

Essentially, the above equation implies that the real price of housing is related to the discounted value of imputed rents.

The World Bank's International Comparison Program (ICP) has provided comprehensive guidelines for the user cost method to calculate rents for owner-occupied housing. For a full discussion and analysis, refer to the table in the report and associated description of the individual items (Table 6.1).[2]

Housing tenure choices

As we have seen in the previous discussion on the user cost, tenure choice hinges on trade-off between owning and renting. We also find a part-own-part-rent option in a few countries, which we discuss later in the chapter. Let us look at tenure choices globally and within the UK compared with other European countries.

Table 6.2 and Figure 6.1 show tenure patterns across a cross-section of countries. As evident, owning-renting share is very different. At the top, Romania, Croatia, Lithuania and Bulgaria have more than 80% share for homeownership, with a very small private renting sector (PRS). Another notable aspect is a vast majority of homeowners in these countries own their homes outright and mortgaged properties are not very common. Mediterranean countries like Greece and Italy also have very high homeownership rates.

The United States, the United Kingdom and Australia show similar homeownership rates with a significant share for homes with some mortgage outstanding. The United States has a higher share of properties with an outstanding mortgage compared to many other countries.

---

2 See Table 1 pp. 9–11 in http://siteresources.worldbank.org/ICPINT/Resources/270056-1255977007108/6483550-1257349667891/6544465-1272763721734/03.03_ICP-TAG03_Guidelines-N.pdf

Table 6.1 Guidelines for the user cost method

Cost categories	Cost item no.	Description of the cost item
Intermediate consumption	UC 01	Expenditure on maintenance and repair of owner-occupied dwellings;
		Expenditures on maintenance and repair are expenditures on replacing or repairing parts of the dwelling that are broken or dilapidated; repairing the roof, replacing window frames, painting the outside of the building are examples. Maintenance and repair expenditures do not extend the service lives of dwellings beyond their previously expected lifetimes and do not involve enlarging the dwelling. Information about expenditures on maintenance and repairs is usually obtained from a household expenditure survey although some countries estimate them from a supply/use table. In some countries expenditures on maintenance and repair of dwellings are incorrectly shown as a separate component of final consumption expenditure of households. When the user cost method is used, they must be included as part of rents and not as a separate expenditure item. Note also that when countries use the standard procedure, rents will already include these expenditures and showing them as a separate item of household consumption expenditure will lead to double counting.
	UC 02	Gross insurance premiums paid on owner-occupied dwellings;
		Gross insurance premiums on dwellings should only include insurance on the dwellings themselves and not on their contents; premiums for the latter are a separate item of household final consumption expenditure. When data are available only for the total of both kinds of insurance, the necessary split between the two can be estimated as being proportional to the relative values of the stock of dwellings and the contents.
	UC 03	Insurance claims paid to owners (minus)
	UC 04	Net insurance premiums paid by owners. (UC 02) − (UC 03)
	UC 05	Total intermediate consumption. (UC 01) + (UC 04)
Other taxes on production	UC 06	Taxes paid by owners on dwelling services:
		Some countries charge taxes on the imputed value of the dwelling services that individuals derive from owning the dwellings they reside in. Taxes on dwelling services are the value of any such taxes. Any subsidies that owner-occupiers receive to assist them in paying current housing expenses, such as government subsidisation of mortgage payments, should be included here as negative taxes
	UC 07	Taxes paid by owners on the value of owner-occupied dwellings and their associated land:
		Taxes on dwellings and land are taxes paid on the value of the dwelling units themselves and the land on which they are located. These taxes are often called 'property taxes'
	UC 08	Total taxes paid by owners. (UC 06) + (UC 08)

(*Continued*)

Table 6.1 Continued

Cost categories	Cost item no.	Description of the cost item
Consumption of fixed capital	UC 09	Consumption of fixed capital on owner-occupied dwellings at current prices (excluding land): *Consumption of fixed capital on the stock of owner-occupied dwellings is measured at current prices and is sometimes called depreciation at current replacement cost. Estimates of consumption of fixed capital should be obtained from estimates of the stock of owner-occupied dwellings valued in current prices. The stock estimates are preferably obtained by the Perpetual Inventory Method (PIM) which is described in detail in the OECD Manual, Measuring Capital: Measurement of Capital Stocks, Consumption of Fixed capital and Capital Services (Paris, 2001). However, many countries that do not have sufficient data to apply the PIM and Table 2 (in the World Bank report) is a worksheet that gives a method for deriving an approximate estimate of the stock of owner-occupied dwellings that can be used by these countries.*
Net operating surplus	UC 10	Current market value of the stock of owner-occupied dwellings at the beginning of the year (including land)
	UC 11	Current market value of the stock of owner-occupied dwellings at the end of the year (including land): *The value of the stock of owner-occupied dwellings represents the value of the net (or 'depreciated') stock of these dwellings valued at current market prices. Table 1 (reproduced above) assumes that the estimates of the stock of owner-occupied dwellings refer to the end of each year and so successive end-year estimates must be averaged to obtain mid-year estimates. The procedure shown in Table 2 (in the report) produces an estimate of the stock for the middle of the year so that this averaging procedure is not required. Note that the stock of dwellings used here must include the estimated value of the land underlying the buildings. Table 2 (in the report) is a worksheet to calculate both the value of the dwellings themselves and the land on which they are situated.*
	UC 12	Current market value of the stock of owner-occupied dwellings at mid-year (including land) ((UC 10) + (UC 11))/2
	UC 13	Real rate of return on owner-occupied dwellings (including land) in per cent per annum: *The choice of the real rate of return used to calculate the net operating surplus was discussed above, where it was suggested that the real rate of return should be set at 2.5% for all countries.*
	UC 14	Real net operating surplus. (UC13) * (UC12)/ 100: *The net operating surplus of owner-occupied dwellings is calculated by applying the real rate of return to the mid-year, current value of the stock of dwellings.*
Expenditure on owner-occupied dwelling services	UC 15	Expenditure on owner-occupied dwelling services. (UC05) + (UC 08) + (UC09) + (UC14)

Source: Based on information from Table 1 and description of cost items in pp. 9–11 of "Guidelines for the User Cost Method to calculate rents for owner occupied housing", International Comparison Program (ICP), World Bank, 2011.

Table 6.2 Dwelling stock: By tenure, global comparison

	Own outright	Owner with mortgage	Rent (private)	Rent (subsidised)	Other, unknown
Romania	95.5%	0.6%	1.0%	0.9%	2.0%
Croatia	85.8%	3.3%	2.1%	1.2%	7.6%
Lithuania	84.2%	5.7%	1.6%	1.8%	6.6%
Bulgaria	81.5%	2.1%	2.8%	1.6%	12.0%
Hungary	73.8%	14.4%	3.9%	3.5%	4.4%
Poland	71.3%	9.8%	5.1%	1.4%	12.3%
Czech Republic	62.4%	14.1%	17.8%	1.4%	4.3%
Greece	61.9%	10.2%	21.2%	0.4%	6.3%
Mexico	61.2%	10.5%	13.1%		15.2%
Italy	57.6%	14.2%	14.5%	4.0%	9.6%
Chile	51.3%	13.3%	18.6%		16.8%
Spain	49.7%	28.4%	12.4%	2.5%	7.0%
Portugal	43.5%	30.4%	13.0%	4.4%	8.7%
Ireland	41.1%	28.3%	15.2%	12.5%	2.9%
France	38.7%	22.7%	21.4%	14.1%	3.1%
Korea	37.8%	15.8%	38.4%	5.1%	2.8%
Belgium	33.0%	33.1%	23.8%	8.5%	1.7%
United Kingdom	32.6%	30.7%	17.3%	18.3%	1.0%
Australia	32.2%	30.7%	31.3%		5.7%
Canada	28.8%	40.5%	30.7%		0.0%
Germany	26.0%	19.0%	50.3%	4.4%	0.3%
United States	22.9%	40.3%	34.9%		1.9%
Norway	21.9%	54.5%	13.3%	0.7%	9.5%
Denmark	15.0%	38.9%	46.0%		0.1%
Sweden	9.8%	52.3%	36.9%	0.5%	0.5%
Netherlands	9.2%	47.3%	42.9%		0.6%
Switzerland	5.0%	34.8%	55.1%	3.4%	1.7%

Source: Based on OECD calculations based on European Survey on Income and Living Conditions (EU SILC) 2014 except Germany; the Household, Income and Labour Dynamics Survey (HILDA) for Australia (2014); the Survey of Labour and Income Dynamics (SLID) for Canada (2011); Encuesta de Caracterización Socioeconómica Nacional (CASEN) for Chile (2013); the German Socioeconomic Panel (GSOEP) for Germany (2014); the Korean Housing Survey (2014); Encuesta Nacional de Ingresos y Gastos de los Hogares (ENIGH) for Mexico (2014); American Community Survey (ACS) for the United States (2014). For a full list, refer to the source. http://www.oecd.org/social/affordable-housing-database.htm

Given a strong secondary mortgage sector in the US, availing mortgage finance to step into the homeownership ladder is more common in the US.

Some major European countries have a very different tenure pattern with the private renting sector being a very sizeable part of the housing market. Germany, Denmark, Sweden, Netherlands and Switzerland show almost half of the market occupied by the private renting sector. Well-functioning private rented sectors offer a viable and long-term tenure option for the households.

A particular case of interest is the UK's private rented sector, which has seen much changes over the decades. Over the last six decades, the share of the private rented sector has declined heavily and homeownership share has increased.

Figures 6.2 and 6.3 show how the shares have changed over the years. Back in 1961, the share of PRS was almost 32%, which steadily declined until 2002–03 and since then, it has been growing rapidly, partly due to rising house prices (much faster than the income

92 Housing tenure, search and mobility

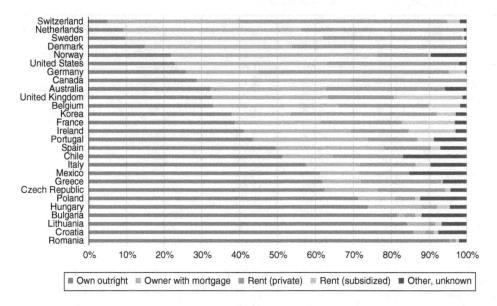

Figure 6.1 Dwelling stock: By tenure, global comparison.

Source: Based on OECD calculations (Table 6.1 above).

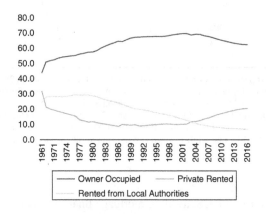

Figure 6.2 Dwelling stock: By tenure, England.

Source: Based on data from MHCLG, UK.

growth) leading to worsening affordability and declining share of the social housing sector.

Several factors have contributed to these changes; rising living standard and preferences, improved mortgage markets with higher level of credit availability, regulations supporting mortgage finance and beneficial tax breaks, and on the other hand, tenants have limited rights with short and uncertain tenancies and rent has been rising faster than incomes. Such situations, coupled with a house price boom, prevented a strong PRS in England. Unfavourable conditions of PRS have led many tenants to opt for homeownership

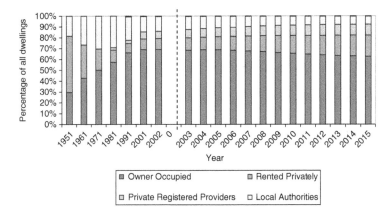

Figure 6.3 Dwelling stock: By tenure, Great Britain.
Source: Based on data from MHCLG, UK.

at any costs. Moreover, buy-to-let schemes and several risks in the sector have driven out the corporate landlords. Many small landlords are operating in the market. This does not allow economies of scale. Big institutions and corporations have left the private rented sector due to a multitude of issues:

- Few economies of scale: except in niche markets, e.g. student, health services, senior/retirement housing, there is not much economies of scale. At the same time, niche sectors need specialised knowledge and experience with high entry barrier.
- Rent stock is often dispersed which makes it difficult to monitor and increase monitoring costs. Monitoring is easier for small landlords.
- Since World War II, there have been continuous and pervasive government interventions in landlord-tenant relationships with much favourable regulation towards homeownership. The ideology of homeownership has been supported by successive governments. Rent controls have also contributed to the sector dynamics.
- There are significant management issues and a high level of reputation risks.
- There are significant transaction costs and management and maintenance costs.
- Due to short tenancies, there is high turnover and high level of vacancy lead to vacant property premium.
- As an investment class, PRS is illiquid and thin market.

Within England and Wales, there are notable variations in housing tenure patterns across regions. Table 6.3 shows eight tenure categories from 2011 Census across various regions. First is the category of 'owned outright'. These households do not have any mortgage obligation left and they own property outright. It can be seen as the category with many elderly and near-retirement homeowners. The second category is 'owned with a mortgage or loan'. These households still have some mortgage obligations left. It is not clear how much obligation is left and this is a mix of households with varying amount of loans still to be paid off. The third category is 'private landlord or letting agency'. This is the core PRS sub-sector with market-driven renting. The fourth category is the core public housing sub-sector – 'rented from council (Local Authority)'. This is the most important source of public housing. Next categories

Table 6.3 Household tenure status, England and Wales Census 2011

Area name	Owned: Owned outright	Owned: Owned with a mortgage or loan	Private rented: Private landlord or letting agency	Social rented: Rented from council (local authority)	Shared ownership (part owned and part rented)	Private rented: Other	Living rent free	Social rented: Other
England and Wales	30.8	32.7	15.3	9.4	0.8	1.4	1.4	8.2
South West	35.4	32.0	15.2	5.8	0.8	1.9	1.4	7.5
Wales	35.4	32.0	12.7	9.8	0.3	1.5	1.6	6.6
East	32.9	34.7	13.3	7.8	0.7	1.4	1.3	7.9
East Midlands	32.8	34.5	13.6	10.1	0.7	1.3	1.3	5.7
South East	32.5	35.1	14.7	5.8	1.1	1.6	1.3	7.9
West Midlands	32.3	32.6	12.8	10.9	0.7	1.2	1.5	8.1
North West	31.0	33.5	14.1	7.7	0.5	1.3	1.3	10.6
Yorkshire and Humber	30.6	33.5	14.4	12.3	0.4	1.5	1.5	5.8
England	30.6	32.8	15.4	9.4	0.8	1.4	1.3	8.3
North East	28.6	33.2	12.4	14.8	0.4	1.3	1.2	8.1
London	21.1	27.1	23.7	13.5	1.3	1.3	1.3	10.6

Source: Based on data from the ONS, UK.

Housing tenure, search and mobility 95

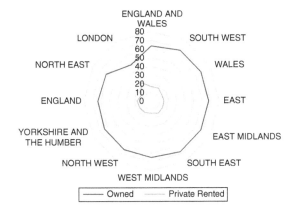

Figure 6.4 Household tenure status, England and Wales Census 2011 (% owned; private rented).
Source: Based on data from the ONS, UK.

are shared ownership (part owned and part rented), other private rented, living rent free and other social rented.

Figure 6.4 shows a very interesting pattern of variation in tenure choice across the regions in England and Wales. London has the lowest homeownership rate in the country with less than 50% share, while all other regions have more than 60% share for the ownership sector. South East and East have the highest rate of homeownership – almost 68%. London has the highest share for the PRS sector – almost 24%. This is due to high house prices and lifestyle choices in London.

Having a healthy, well-functioning PRS sector is important as it supports the renting needs of households who cannot afford ownership, and moreover, it supports a key stage of the housing life cycle. A healthy PRS can also bring stability to the ownership sector with less demand and better quality of financially able households demanding ownership properties. As we have seen in the international comparison, countries like Germany, Denmark and Netherlands have a strong, well-functioning PRS sector that lends much-needed support to housing market stability. IPPR (Institute for Public Policy Research), a progressive think tank, have compared UK and Germany housing markets through research studies[3]. These two markets are quite different – in Germany, PRS holds almost 40% of the market, while in England is it only about 19%. The report finds a number of similarities between the German and English rental markets in terms of the search processes, tenants' background checks and high level of deposits. Both countries have landlords who operate small number of properties.

However, it also notes many areas of divergence, in which the German PRS seems to provide more security to the tenants.

- Most tenancies are indefinite and it is very difficult and rare for landlords to evict tenants unless there are compelling reasons.

---

3 Davies, B., Snelling, C., Turner, E. and Marquardt, S. (2017) Lessons from Germany: Tenant power in the rental market. IPPR. http://www.ippr.org/publications/lessonsfrom-germany-tenant-power-in-the-rental-market

- Typically, in Germany tenants pay 3 months rent as deposit which must be invested in a savings bank account. In England, it is about 1–2 months rent and must be deposited with the national tenancy deposit scheme.
- The assured shorthold tenancy (ASTs) is about 6–12 months in typical contracts. Tenants can be evicted at the end of the initial contract period without any reasons given and with a notice period of 1–2 months.
- However, tenants in Germany enjoy a lot more stability and assurance in their chosen tenure. As a result, the turnover ratio is much lower, and tenants tend to move house a lot less frequently than those in England. German tenancies are, on average, almost 11 years compared to only about 2.5 years in England.
- In Germany, only about 23% of tenants pay more than 40% of their income towards housing cost, compared to approximately 33% in England.
- German tenants typically spend only a quarter of their incomes, while in England, tenants spend almost 40% of their income.
- More important, Germany has strong rent control restrictions in terms of in-tenancy rent increases. There is effectively no rent control system in England.
- In terms of political economy, German tenants provide a significant political powerbase and can thus lobby for favourable regulatory support much more effectively.

In terms of recommendations to make PRS function better in England, IPPR recommends several steps.

- Greater security of tenure with longer-term rental contracts and more protection for tenants.
- Government-backed support and local authority-led schemes to allow longer tenancy.
- New build-to-rent homes should be encouraged.
- Some form of rent control.
- More granular information about the stock and applicable rent from the government agencies.
- Local government should work with the insurance industry to establish an insurance product to replace the tenancy deposit.
- More political power and activism to help tenants' association to voice concerns and lobby for better market and living conditions.

The differences in tenure patterns across the regions in England and Wales can be partly explained in an affordability situation. A simple, quick yet useful indicator for affordability is the income multiple. It is obtained by taking the ratio of house prices and income. Typically, average, median or lower quartile measures are used to calculate the ratios in a region or geographic area. While the average may suffer from biases due to extreme values (very high or very low figures), median and quartile measures based on the ranked data can be less prone to such biases. This income multiple is an important indicator. From the mortgage availability and credit worthiness point of view, often lending institutions such as banks, building societies use 4 to 4.5 as an acceptable income multiple to approving mortgage amount. So, a household earning £50,000 of gross annual income may be allowed to take maximum £225,000 as mortgaged or loan amount. This is a thumb-rule and there are several other factors including credit rating, past credit behaviour, any other ongoing loan items (such as car loan, any other loans), monthly expenditure items and personal obligations (such as alimony, maintenance) are taken into account to derive the maximum mortgage amount that an

Table 6.4 Affordability across England and Wales (ratio of lower quartile house price to lower quartile earnings)

	1997	2001	2003	2005	2007	2009	2011
England	3.6	4.1	5.2	6.8	7.3	6.3	6.5
North East	2.9	2.6	3.1	4.8	5.5	4.9	4.6
North West	3.0	2.9	3.3	5.0	5.9	5.0	5.0
Yorkshire and Humber	3.1	3.0	3.5	5.3	6.3	5.2	5.1
East Midlands	3.2	3.6	4.9	6.5	7.0	5.7	5.7
West Midlands	3.5	3.7	5.0	6.5	6.9	5.8	6.0
East	3.7	4.8	6.7	8.0	8.6	7.2	7.6
London	3.9	6.0	7.7	8.5	9.1	8.0	9.0
South East	4.2	5.6	7.5	8.6	8.9	7.7	8.2
South West	4.0	5.2	7.1	8.6	8.9	7.6	7.8

Source: Based on data from MHCLG, UK. http://www.communities.gov.uk/documents/housing/xls/152924.xls

individual or a household can take on. However, the income multiple provides a very good indicator of someone's affordability. Table 6.4 provides income multiples, based on lower quartile measures of house prices and incomes, across regions in England and Wales. This is also available by local authority by the government. Such figures can also be obtained in many parts of the world. This is not difficult to calculate. Countries do differ in income multiples. There will obviously be significant variation within the country. In big countries like the US, India, China, Brazil or Russia, the income multiple can vary markedly. The income multiple also varies significantly over time due to changes in house prices and income levels. One of the key concerns, globally, is the rising divide between house price inflation and income growth. In some cases, while house prices have increased, the income level has stayed largely flat or increased only slightly. This has resulted in the income multiple rising over the years. Especially during the run-up to the 2006–07 real estate boom, the income multiple rose significantly within a short span of time (about 3–4 years).

As Table 6.4 shows, there has been significant variation across the regions in England and Wales. Overall in England, the income multiple almost doubled in 10 years – 3.57 in 1997 to 7.25 in 2007. In some areas, it has increased by more than 2.25 times. Recall that lending institutions may use 4.0–4.5 as the affordable income multiple. In that case, besides the North East, no other regions come close to those affordable levels in 2011. If we take London, even at the lower quartile level, it stands almost 9 times the income level. This shows the severity of unaffordability in England and Wales. It is also remarkable how the income multiples have changed so much in a short time period and every 2 years.

Shared ownership[4]

In England, as evident from Table 6.4, the affordability problem is severe in Greater London and the South-East England region. If a household wants to buy a house in London, at the lower quartile level, that would be almost 9 times the income, i.e. no mortgage lender will be willing to allow so high an income multiple. Unless the non-mortgaged portion is financed by personal savings or inheritance or help from friends and families, it is not possible to buy a house in London, South East, South West or in the East for a large number of households with

---

4 Much of these discussions is based on Nanda and Parker (2011, 2013, 2015).

low income levels. The shared ownership (SO) model is an intermediate housing mechanism, where the idea is to enable households facing severe affordability constraints, such as in London, to purchase a part of the house (as per the financial condition and typically as low as 25% of the total equity) and the rest is rented. So, the monthly housing cost would have two major components – mortgage obligation on ownership share and rental payment on the rest. This model, in theory, allows many households in regions suffering from a severe affordability issue to get into the homeownership ladder. One of the crucial aspects of the SO model is that the households would keep on increasing their equity share as their financial conditions improve, i.e. income grows. This is called 'stair-casing'. After a certain number of years and depending on the stair-casing rate, a household would be able to own 100% of the equity.

Table 6.5 shows housing deprivation across several countries – Romania, Mexico, Bulgaria, Latvia and Hungary. The developed countries with high per capita income such as the UK, Germany, Iceland, Ireland, Luxembourg, Netherlands, Norway, Sweden and Switzerland show very low or nil housing deprivation. However, in several pockets of low income and high house price areas within the developed countries, the housing deprivation can still exist. Policymaking units often need to think several ways to close the housing affordability issues, beyond just creating opportunities for new supply. In a small number of countries, intermediate housing mechanisms have been seen as a part of the menu of solutions for affordable housing problems. Elsinga (2005) and Scanlon and Kochan (2011) have provided detailed discussions of these schemes; for example, England (Homebuy/Shared Ownership), the US (Limited Equity Housing Cooperative – LEHC), Finland (the Right of Occupancy) and The Netherlands (Protected Homeownership - Beschut Starterslening / Eigenwoningbezit/ Koophuur). Two main types of alternative tenure have been highlighted: subsidised forms of homeownership (e.g. Homebuy in England and Starterslening in the Netherlands), which can be temporary tenures with the objective of encouraging users to become full homeowners, and the second type of tenure is a permanent hybrid with a bundle of rights and duties related to renting and owning. Elsinga (2005) also noted key distinctions between partial ownership arrangements (i.e. the Right of Occupancy in Finland and Koophuur in the Netherlands) and regulated full-ownership arrangements (i.e. Limited Equity Cooperatives in the United States and Community Linked Ownership (Maatschappelijk Gebonden Eigendom – MGE) in the Netherlands). Shared Ownership (SO) in England may as well be creating a tenure type that is at the intersection of these tenures (Elsinga, 2005: pp. 84–85).

With various forms of intermediate housing tenure options, a significant factor that differs among the country case studies, Nanda and Parker (2015) argue, is the degree of local authority, autonomy and consequential ability to innovate locally. Municipalities in the US, Germany and Australia tend to have more independence than those in the UK in terms of policy and budgetary matters. As Nanda and Parker (2015) noted:

> solutions also differ based on market dynamics and where demand subsidies are inexpensive, but these may not actually solve affordability issues in the long run. Supply subsidies are also seen as useful, but come with political considerations and limits (e.g. which groups to target?). The questions raised by the UK experience therefore appear somewhat universal in terms of difficulties in defining the market, identifying and reaching target groups and problems with the recycling of units and associated subsidy.

McKee (2010) analysed the Scottish housing policy of using tenure-mix as a regeneration strategy through the Low-Cost Initiatives for First Time Buyers (LIFT) strategy. This mix included shared equity and shared ownership schemes. Tenants' survey revealed that there

Table 6.5 Housing deprived population across the income distribution *(2014 or latest year available; Share of deprived population, bottom and third quintile of the income distribution, in per cent)*

	Bottom quintile	3rd quintile
Romania	45.6%	10.5%
Mexico	41.1%	23.2%
Bulgaria	28.6%	4.7%
Latvia	18.0%	5.9%
Hungary	15.0%	1.3%
Lithuania	8.9%	3.4%
Poland	7.1%	1.5%
Croatia	4.1%	0.1%
Slovak Republic	2.9%	0.2%
Austria	1.8%	0.4%
Chile	1.8%	0.6%
Estonia	1.3%	0.8%
Czech Republic	0.9%	0.2%
Slovenia	0.9%	0.0%
Belgium	0.8%	0.0%
Denmark	0.8%	0.0%
Portugal	0.8%	0.2%
France	0.4%	0.0%
Greece	0.4%	0.2%
Korea	0.3%	0.0%
Finland	0.2%	0.1%
Spain	0.2%	0.0%
Italy	0.1%	0.1%
United States	0.1%	0.0%
Cyprus (a, b)	0.1%	0.0%
Malta	0.1%	0.0%
United Kingdom	0.0%	0.1%
Germany	0.0%	0.0%
Iceland	0.0%	0.0%
Ireland	0.0%	0.0%
Luxembourg	0.0%	0.0%
Netherlands	0.0%	0.0%
Norway	0.0%	0.0%
Sweden	0.0%	0.0%
Switzerland	0.0%	0.0%

Source: Based on OECD calculations based on European Survey on Income and Living Conditions (EU SILC) 2014 except Germany; Encuesta de Caracterización Socioeconómica Nacional (CASEN) for Chile (2013); the German Socioeconomic Panel (GSOEP) for Germany (2014); the Korean Housing Survey (2014); Encuesta Nacional de Ingresos y Gastos de los Hogares (ENIGH) for Mexico (2014); American Community Survey (ACS) for the United States (2014). For a full list, refer to the source.

are significant questions on ability and conducive conditions that would allow and encourage tenants to stair-case at a regular interval as their financial conditions improve and become 'full' homeowners. As pointed out by Elsinga (2005), these differences may be driven by a variety of motivations, backgrounds and ideologies in different countries. For example, in England to a large extent, the development of the Shared Ownership and Homebuy programmes are part of strategic housing policies across the regions designed to solve affordability issues, especially in areas with a very high level of house prices. However, these can still be categorised and justifiably viewed as schemes and vehicles through which the ultimate goal seem to be attainment of homeownership. The stated objective is to make properties available to those who are otherwise sandwiched between social housing, private renting and the private owner-occupation markets.

The Finnish Right of Occupancy scheme was designed as an alternative to subsidised rental units in which the occupiers take stakes in the capital. The motivation of the Low Equity Housing Cooperative (LEHC) approach in the United States may actually be in response to problems of neglected property and one way for local authorities to utilise heavily-depreciated housing stock in a more productive manner. Following these policy prescriptions by the governments, two main types of new tenures have cropped up: a 'social ownership sector' and a 'leg-up to full-ownership' approach. However, these alternative tenures are fraught with many challenges and Elsinga (2005: p. 92) points out that:

> it is by no means easy to introduce into the market a product that is neither a rental property nor an owner-occupied property. If occupiers are unable to claim tax relief or rent rebates, the housing tenure will quickly lose its attractiveness. Moreover, lower-income consumers face a risk of losing their housing allowances. For this reason, attempts to combine the best of both worlds often result in the worst of both worlds.

Interestingly, Elsinga also notes that such new housing tenures are not welcomed by the lenders, developers and other stakeholders in the Netherlands, as well as in other countries. This could be a reason why the market share for these alternative structures has remained relatively modest, being only 0.4% of stock in England, 0.4% in the US and 0.05% in The Netherlands by 2005.

Under shared ownership in England, it is expected that the owner occupier will start with a portion (starting around 25% or more) and more importantly, the owner occupier is expected to increase that share over time, i.e. 'stair-casing' up to full-ownership (Clarke et al, 2008; Ford, 2006; Wallace, 2008). However, Nanda and Parker (2015) argued that little empirical confirmation of this process currently exists. With several variations of schemes and subsidies, the SO offerings have been refined and yet the take-up is low. Generally across all the schemes, buying a portion, typically 25–75%, of the house value is the key feature.

Nanda and Parker (2015) analysed with an example of how shared ownership can solve the affordability problem for a prospective homeowner. The example provides the commonly cited economic rationale for the intermediate housing mechanism (such as SO). The economic rationale is frequently argued on the fact that a large number of households experience sizeable deposit constraints and are often priced out of the market due to ever increasing house prices coupled with stagnant income levels and these households need to be supported by affordable housing as well as market-oriented schemes. By purchasing only a part of the total housing equity (in expensive areas), the deposit constraint is not restrictive and a lot more within the reach in achieving homeownership dreams. Let's take an example – the average price for a two bedroom house in London is £400,000. If the householder applies for a loan, a lender in general would be willing to lend no more than 80% of the value. This implies the rest of the house value, i.e. 20% of the house value or £80,000, would need to be sourced from somewhere else. This can come from past savings, wealth, inheritance or help from friends and families. This deposit constraint can prove to be exorbitant for many households whose income level does not allow for a high rate of savings or any savings on a regular basis. By taking advantage of an SO scheme, a household can get around these constraints. Suppose the household buys 25% of a housing under the SO scheme. As a result of this, the deposit constraint is much lower – only £20,000. This means that the household can now get on the housing ladder. However, two major issues that can derail this argument, as pointed out in the literature and by Nanda and Parker (2015): first, income has been mostly stagnant or growing at a much lower rate than house price inflation and general consumer prices, and second, the total cost of housing will need to include all cost components occupying the housing unit under the SO scheme. These cost items may include opportunity cost or interest income lost on the initial deposit of £20,000, mortgage payment on 25% of the house value, rental applicable on 75% of the house value, and service fees, maintenance fees and any management fees. In total, all these can be quite significant. The question is: does this work out in favour of this ownership option within the recognised User Cost of Housing framework as discussed in the previous section. Elsinga (2005: pp. 89, Figure 5.1) shows that the expenses of an owner-occupier could be higher than those of a renter within the user cost comparison across four tenure types.

While the economic argument may not quite work in reality, the political and ideological rationales are supported strongly. The SO has been widely viewed as part of a policy framework driven partly by an ideological bias supporting homeownership. Homeownership is generally seen as the most favoured housing occupation. Homeownership is good for the households as it has been a great vehicle for creating wealth over the decades. It can be beneficial for the family as it can bring stability to family life and it is good for the neighbourhood, region and country. It is also seen as good for the economy due to the possibility that housing wealth can boost spending capacity and may add to aggregate expenditure. Such rationales may have driven governments to support and enhance the offering and allocate valuable subsidies and this is also the case in other countries (Gurney, 1999; Elsinga, 2005). Many studies have analysed and documented the ideology of homeownership such as Kemeny (1981), Saunders (1984), Forrest (1987) and Murie (1997). Gurney (1999) argued that support from the successive UK governments and associated direction to the policymaking have fuelled private development and homeownership being viewed as a social 'good'. However, this came at the cost of marginalisation of other housing tenures such as PRS. Nanda and Parker (2011) also noted that motives backed by the above rationales have led to policies which, in turn, meant 'socio-tenurial polarization', a continued residualisation of public housing and an associated normalisation of private-sector driven homeownership. Various intermediate housing models have been designed to supply private-sector led 'affordable housing' over the past three decades and by some to promote 'mixed communities' (for example; Berube, 2005; Goodchild and Cole, 2001).

Wallace calls for deeper examination of the SO, stating that:

> understanding the dynamics of this growing sector is imperative for investment and business planning reasons as well as in establishing the outcomes for low to middle income home-owners. Studies have mostly examined the potential demand for LCHO and the owners' satisfaction with the products after purchase, but the longer-term outcomes for owners have not previously been a focus of research.
>
> (Wallace, 2008: p. 74)

Furthermore, some studies in the literature also raised concerns related to the questions of mobility across tenures and housing markets. The cost aspects as noted above can be prohibitive and a significant deterrent for stair-casing potential. One key point that Nanda and Parker (2015) raised is whether the SO is *de facto* a temporary or a transitional tenure. Overall, while it may be appealing on theory, there are significant concerns and practical issues as mentioned above on whether SO can really achieve what it is supposed to achieve.

Table 6.6 shows various schemes of intermediate housing mechanisms in England.

Table 6.6 Basic overview of intermediate housing (SO) schemes in England

Intermediate housing scheme	Basis	Details
New Build Homebuy (traditionally shared ownership)	Part buy/part rent on new build homes	• Purchase between 25 and 75% at the outset. • Purchase additional shares until 100% is owned - 'staircasing'. • Eligibility – new Build HomeBuy is usually available to first-time buyers with a maximum income defined by the scheme provider • Also targeted at 'priority groups' such as existing social housing tenants, people on local authority waiting lists and key workers.
Homebuy Direct	Shared equity scheme	• Both builders and Government provide funding for a potentially sizable deposit on a new home in the form an equity loan (shared equity). • The scheme is available for new build properties. • The purchaser takes out a mortgage to cover at least 70% of the purchase price (up to a maximum of 85%). • This is topped up with an equity loan covering up to 30% of the price. The Government shares the cost of the equity loan 50/50 with the developer. • The maximum value of homes purchased through HomeBuy Direct is £300,000.
Rent to Homebuy	Potential homebuyers rent at a discounted rate before they purchase	• Potential SO home buyers pay a reduced rent on a new-build home. • Tenants can then save their discount towards a deposit for a SO property purchase. • The tenant pays a rent set at no more than 80% of the current market rent on a home for up to 5 years. • Enables people who might otherwise have trouble saving a deposit for an affordable mortgage to take part in Rent to HomeBuy, which could eventually lead to purchasing a SO home. • For new build properties the eventual purchase would take place under the new build Homebuy scheme. It is of particular benefit to RSLs who have a LCHO scheme but finding it difficult to market the units during the current financial climate.
Social Homebuy	Allows existing secure or assured tenants of participating HAs and council housing shared ownership basis or outright, with the benefit of a discount	• Tenants usually buy their home on a shared ownership basis with a minimum share of 25% and outright purchase is also allowed. • The maximum discount available varies from £9,000 to £16,000, depending on the location of the property. • Discounts are also available on any subsequent shares purchased.
Open Market Home Buy* (*to become *FirstBuy*, first homes available Sept. 2011)	Purchase 75% and receive 2 equity loans to cover the remainder.	Purchase 75% of a property and receive two equity loans of 12.5% from the mortgage lender and the Government or Homebuy agent. *(For Open Market HomeBuy, purchase at least 80% of the purchase price and equity loan funding of up to 20% of the purchase price, which is split between HCA and the housebuilder. http://www.homesandcommunities.co.uk/firstbuy)*
Do It Yourself Shared Ownership	Funded by Housing Associations, available for people to buy a share of a property on the open market	• The tenant/owner selects a property they would like to purchase on the open market and approaches the Housing Association to purchase the remaining share (normally based on 75% to 25% share, but can sometimes be negotiated). • It can be used in mortgage rescue.

Source: Based on the information from Annex 3 Nanda and Parker (2011) "Analysis of the Intermediate Housing Market Mechanism in the UK" http://centaur.reading.ac.uk/26952/1/1711.pdf

A 2014 McKinsey Global Institute (MGI) report[5] has shown very alarming figures of unaffordability across the world. The study defines the affordability gap as the difference between the cost of an acceptable standard housing unit (which may vary by location) and what households can afford to pay using no more than 30% of income. The analysis has drawn on MGI's Cityscope database of 2,400 metropolitan areas, as well as case studies from around the world. It found that the affordable housing gap stood at a staggering $650 billion a year and that the problem would only grow as urban populations expand, i.e. when almost two-thirds of the world's population would live in urban areas in 2050. The issue, if unchecked, would be enormous. The study further estimated that there could be 106 million more low-income urban households by 2025. Another alarming figure, the study notes, is that to replace the inadequate housing and build the additional units required to close the gap by 2025 would require $9–11 trillion in construction spending alone. With land values taken into account, the total cost could be as much as $16 trillion, out of which almost $1–3 trillion would need to be funded by the public purse. The study identified four approaches, if used together, could potentially reduce the cost of affordable housing by 20 to 50% and substantially narrow the affordable housing gap by 2025. Those solutions are all market-based – lowering the cost of land, construction, operations and maintenance, and financing – could make housing affordable for households earning 50 to 80% of median income.

Let's have a look at the social rental housing provision across a selection of countries. Table 6.7 provides a summary. The OECD Questionnaire on Social and Affordable Housing (QuASH), 2016 and 2014, is the basis for this summary. For a fuller description of various aspects and information on all available countries, refer to the survey from OECD. Countries vary in terms of importance and criteria places on income threshold, household composition or size and housing situation such as homeownership.

The McKinsey study further notes that

> the successful application of these approaches depends on creating an appropriate delivery platform for housing in each city. Policymakers, working with the private sector and local communities, need to set clear aspirations for housing throughout their cities. Critically, a minimum-standard housing unit must be defined in each of them. But an excessively ambitious minimum can discourage the construction of affordable homes and force more low-income households into informal housing. A better solution is to set standards that reflect rising aspirations – a housing "ladder" that can start with something very basic that might, for example, have communal kitchens and baths and serve as transitional housing for new arrivals.

Like this study suggests, there is no one solution. Both public and private sectors need to come together to have any meaningful change to this scenario. Otherwise, there can be a very significant inequality in terms of housing outcome, which can also lead to several societal issues and economic challenges.

Table 6.8 shows the share of social rental housing stock in 2000 and 2015 across a selection of countries.

The governments will need to spend much more than what they are spending to be able to close the housing deprivation. Figure 6.5 shows that the governments are not spending much at all in supporting social rental housing.

---

5 https://www.mckinsey.com/featured-insights/urbanization/tackling-the-worlds-affordable-housing-challenge

Table 6.7 Summary of social rental housing across selected countries: Criteria assessed in selecting eligible households

	Year	Definition	Summary overview	Criteria for selecting eligible households
Australia	2015	Public housing (including State and Territory public housing, and State Owned and Managed Indigenous Housing), Community housing, and Indigenous community housing.	Rental housing provided by not-for-profit, non-government or government organisations to assist low to moderate income people and families who are unable to access suitable accommodation in the private rental market.	Ability to sustain a tenancy; no criteria on household composition; must not own a property
Canada	2014	Social housing	Social housing is subsidised housing that is usually targeted to low- to moderate-income households who would otherwise be unable to afford suitable and adequate housing. It is typically owned by governments, non-profit groups or co-operatives.	Takes into account income criteria
France	2015	Social housing or moderate rent housing (*Habitation a loyer modere*, or HLM)	A social dwelling is housing 1) covered by a contract with the State opening right to personal accommodation help, rented as the main residence, at a reduced rent; the maximum rate varies according to the type of funding (PLUS, PLAI, PLS) and the geographic area where it belongs; 2) for low or modest income households; 3) funded by state subsidies; 4) administered by a social landlord.	Income thresholds vary across the different subsidy schemes; do not take into account household composition or housing situation
Germany	2014	Subsidised housing or social housing promotion	Subsidies are provided by the federal states in exchange for the use of a dwelling for social purposes (enforcing income ceilings and lower rents) for a period of 20 to 40 years, depending on the funding programme. All kind of providers are eligible for subsidies (municipalities, cooperatives, private landlords, commercial developers and investors with a variety of shareholders).	Criteria vary across Lander and different subsidy programmes; takes into account income and household composition.
Japan	2015	Public housing	Public rental housing for low-income households, the elderly, the handicapped and households with children.	Up to 25% of household income distribution, up to 40% for the elderly and household with children; takes into account income and household composition.

New Zealand	2015	Housing which belongs to Housing New Zealand Corporation (HNZC) and community housing.	Premises receiving public subsidies, let by or on behalf of a registered community housing provider or by the public company Housing New Zealand Corporation, with income-based rents	Takes into account income and property ownership
United Kingdom	2015	Social rental housing/ affordable rental housing	The term social housing refers to dwellings let by local authorities and housing associations (registered social landlords), provided to specified eligible households whose needs are not met by the market. Since local authorities stopped building homes in large numbers, non-profit making housing associations are mainly responsible for building new social housing. Since 2011, they are encouraged to build properties with affordable rent (up to 80% of market rents) and at fixed term tenancy, instead of the formerly typical social rents (usually half the market rate) with lifelong tenancies.	Application is open to all British citizen or a citizens who have the right to stay in the UK for an unlimited time;
United States	2015	Public housing and supportive housing	Public housing is direct provision of rental housing by the states and local housing agencies with subsidies form federal government. Furthermore, the federal government provides subsidies to private entities (both for profit and non profit) who own and manage supportive housing for elderlly and disabled.	Up to 80% of local area median income.; 40% of new admissions for each local agency must be for those below 30% of local area median income

Sources: OECD Questionnaire on Social and Affordable Housing (QuASH), 2016 and 2014. Compiled and adapted from http://www.oecd.org/social/affordable-housing-database.htm

Housing mobility

Housing mobility has been a key area of research. Issues dealing with drivers of mobility, policies that can support mobility positively, socio-economic and geographic heterogeneity with respect to mobility are key. Dieleman (2001) discussed residential mobility in detail. Several aspects have been highlighted:

- A residential move typically involves joint decision-making by members of a household. Such joint decision-making can give rise to intra-household bargaining in terms of locational attributes, physical attributes of the house, funding structure, etc. If the financing involves more than one member of the household, it can also become a subject of much deliberation in terms of ownership, bequeath issues, etc.
- There is a very strong link between place of residence and place of work. Much research have been conducted on examining this link at various spatial scales. Neighbourhood networking and its influence on the link between residence and workplace have been studied (see Bayer et al., 2008). The relevance of social networks and local interactions

Table 6.8 Total and social rental housing stock across selected countries

	2000		2015	
	Total number	% of total housing stock	Total number	% of total housing stock
Australia	422,189	6.5	421,233	4.9
Austria	660,851	22.6	890,000	26.2
Canada	640,800	5.2	..	..
Czech Republic	3,149	0.1	21,658	..
Finland	371,607	16.2	334,666	12.8
France	4,735,000	19.1	5,398,000	18.7
Germany	2,570,605	7.2	..	..
Japan	1,729,200	3.9	..	..
Netherlands	2,440,000	36.7	2,481,000	34.1
New Zealand	85,000	6.2	..	..
Norway	73,704	3.7	105,567	4.6
Poland	1,896,000	16.0	1,163,000	8.3
United Kingdom	5,394,000	21.3	4,954,000	17.6
United States	..	..	5,038,578	4.3

Source: Based on the OECD Questionnaire on Affordable and Social Housing, 2016; for Denmark and Ireland, information on social rental housing in the year 2000 is taken from Dol, K. and Haffner, M. (2010) Housing Statistics in the European Union, Ministry of the Interior and Kingdom Relations, The Hague. For a full list, refer to the source.

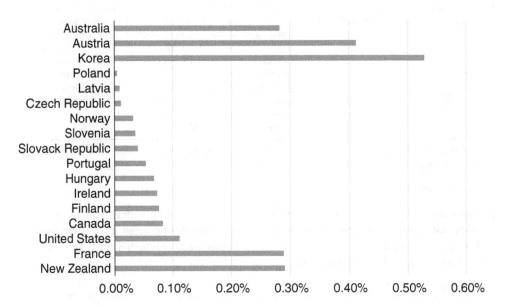

Figure 6.5 Public spending on supporting social rental housing in selected OECD countries (% of GDP).

Source: Based on OECD Questionnaire on Affordable and Social Housing (2014, 2016); for a full list, refer to the *source*.

for economic outcomes has been examined with detection and measurement of social interactions at the level of the neighbourhood level. Bayer et al. (2008) argue that the proper identification of such neighbourhood effects is complicated due to non-random sorting of households into territorial units and unobserved individual and neighbourhood attributes can mar our ability to isolate the effects with correlation in unobservables adding severe biases. The results, using data from the Boston Metropolitan Area, show the existence of significant social interactions at the block level (lowest unit of US Census data) – residing on the same versus nearby blocks increases the probability of working together by over 33%. As a consequence, individuals are about 6.9 percentage points more likely to work with at least one person from their block of residence than they would be in the absence of referrals. This result is robust to the possibility of sorting into specific blocks and reverse causation (i.e. the idea referrals from the workplace to suggest residential opportunities).

- There are significant issues related to challenges of finding an alternative dwelling if the most preferred house is unavailable. It also concerns how households are matched to houses. There is extensive information on the household attributes, the household life course, and the educational and job career which determine the propensity to move and the choice of a dwelling (for reviews, see Clark and Dieleman, 1996; Dieleman and Mulder, 2002; Strassmann, 1991, 2000, 2001).
- The major elements of the 'housing bundle' – the attributes of dwellings households consider when deciding whether to move or which dwelling to choose – are also well researched (Molin et al., 1996).
- Actual choice ('revealed preference') and stated preference are two approaches that have been developed to analyse the residential mobility process at the micro level (Mulder, 1996).

Three well-documented regularities are particularly prevalent (Dieleman, 2001).

1 **There is strong correlation between the rate of mobility and the stage in the life cycle of a person.** In all developed societies, young adults between the ages of 20 and 35 are by far the most mobile segments of the population. However, with ageing population rising across the world creating a huge demand for retirement housing and with developers providing suitable retirement housing supply, the move to retirement housing is becoming significant.
2 **There is strong correlation between the rate of residential mobility and the size and tenure of the present dwelling.** Households in relatively large units are less mobile. This could be due to the fact that there is probably not much 'room stress' compared to the households in smaller dwellings, and owner-occupiers have a much lower mobility rate than renters, although mobility rate among renters vary significantly across the countries.
3 **There are interrelationships between the housing career of a person or household and events in other domains of the life course**, such as family formation and dissolution, and completion of educational attainment, start of employment and other life events. These work as significant triggers for residential mobility decision (Mulder and Hooimeijer, 1999).

108 *Housing tenure, search and mobility*

Figure 6.6 shows that factors of various territorial units are identified in relation to residential location. There are international, national and sub-national (i.e. regional/metropolitan) drivers. In models of residential mobility, it may be necessary to control for all these factors to be able to derive robust inferences.

One of the key challenges in modelling residential mobility is the treatment of supply side dynamics. This can have a significant bearing on the model specification and identification strategy. Strassmann (2001) points out that European and North American researchers take quite different approaches when analysing the process of residential mobility. European researchers emphasise residential mobility at the micro (i.e. household) level and stress the complexity of the mobility process. Mobility models often treat the supply of housing as an exogenous factor, purportedly because complex government interventions in land use and in finance, construction and pricing of housing constrain the supply of (new) housing. In the United States, dwellings can be designed, financed, built, sold and rented with less government controls. This is reflected in the way many North American researchers approach the process of residential mobility. They give primacy to market forces and economic modelling and supply-side factors are often endogenous to the models.

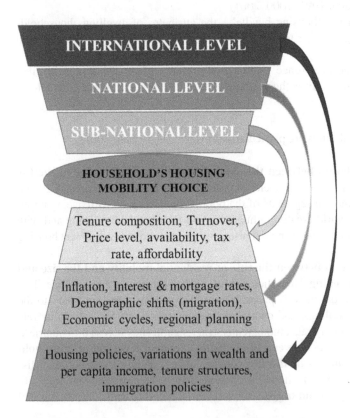

Figure 6.6 Residential relocation and drivers at different scales.
Source: Adapted from Dieleman, 2001.

Housing search process

Recall two unique attributes that we have discussed in Chapter 2 are 'high transaction cost' and 'information asymmetry' in housing transactions. Searching for a home takes time and involves significant monetary and non-monetary costs. Information required to decide on a house is enormous and those can usually be sourced from multiple organisations. Several steps need to be completed in order to successfully transact a house, from both a seller's and buyer's perspectives. Several individuals and organisations are involved, besides the two main parties – buyer and seller. The housing search is often fraught with several uncertainties. Depending on the market condition and geographic area, usually it can take 2–3 months to complete a transaction. Due to the complexities and uncertainties, we often need intermediaries (brokers, licensees, estate agents) to conduct the process on the seller's or buyer's behalf. Let's look at typical journeys of buyers and sellers in a housing transaction.

The Office of Fair Trading (Home buying and selling: A Market Study, 2010) described several steps in the transaction process from buyers' and sellers' perspectives. Figure 6.7 provides a typical journey map for a seller and a buyer in two panels respectively. The following description from the report is edited to keep the discussion relevant to recent updates.

The legal relationship in terms of an agency in real estate brokerage depends mainly on whose side the agent is representing. Table 6.9 provides the definition of types of agency relationships in England.

The US brokerage industry, though similar in several ways, is structured differently. According to the data from their website as of April 9, 2018, the National Association of REALTORS® in the US has over 1.1 million members, 54 state associations (including Guam, Puerto Rico and the Virgin Islands) and more than 1,300 local associations. Licensed agents

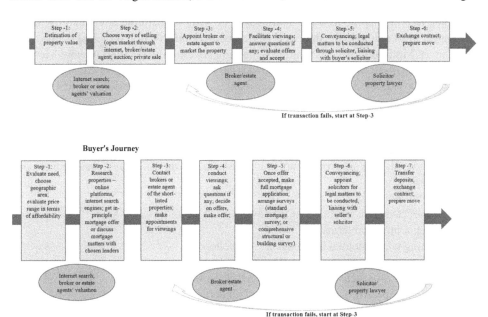

Figure 6.7 Seller's and buyer's journey (based on Office of Fair Trading study, 2010).

Table 6.9 Real estate agency definitions in England

Agreement	Description
Single agency contracts	**There can be two types of contracts under single agency.** **Sole agency contracts:** A single estate agent retains the exclusive right to sell the property for a certain period of time. The seller has to pay commission to the estate agent if contracts are exchanged to a purchaser introduced by the estate agent (or introduced by another agent) in that period or with whom the estate agent had negotiations. However, seller may sell privately without paying commission. However, under **sole selling rights,** estate agent is entitled to commission upon exchange even if the property is sold privately. If contracts exchanged afterwards, commission is payable if the property sold to a purchaser introduced by the estate agent or with whom the estate agent had negotiations during that period.
Joint sole agency contracts	Two or more estate agents act as sole agents and split the commission if one of them is successful.
Multi agency contracts	More than one estate agents are instructed to sell the property at the same time. Commission is paid only to the estate agent that finds a buyer.

Source: Based on and adapted from Table 3.5. Office of Fair Trading (Home buying and selling: A Market Study, 2010).

may join local, state and national associations, and thereby call themselves a 'REALTOR', which is a trade-registered name only available for use by members who agree to adhere to a specified code of conduct and pay local, state and national dues. It is possible to have dual agencies when a real estate broker or salesperson represents the buyer and the seller in the same transaction. Dual agency is usually governed by laws that requires a full disclosure and written informed consent of the dual agency be made to both the buyer and the seller. A dual agent must maintain the duty of confidentiality of information provided by the seller and the buyer. Moreover, real estate professionals are governed by various laws and licensing standards in the US. Brokers in the United States earn a considerably higher commission rate compared to those in the UK. The US Department of Justice (DOJ) lists the typical services that real estate brokers have performed[6]:

- Marketing the home.
    - Marketing services include listing the property in the local multiple listing service (MLS), placing advertisements in local media and on the internet and hosting open houses.
- Reviewing contracts.
    - Contract review might include providing advice on pricing, home inspections or other contractual terms.
- Negotiating with potential home buyers and sellers.
- Locating potential properties for prospective buyers.
- Arranging for prospective buyers to inspect properties.
- Providing prospective buyers and sellers with pertinent information about a community such as relative property values, most recent selling prices and property taxes.

6 See https://www.justice.gov/atr/competition-real-estate-questions-and-answers

- Apprising potential buyers of financing alternatives.
- Assisting in the formation and negotiation of offers, counteroffers and acceptances.
- Assisting with the closing of the transaction.
  - Closing services might include assistance with handling paperwork.

For these efforts as mentioned above, real estate brokers typically charge a single fee (the commission) of almost 5% to 6% of the sales price of the home. Delcoure and Miller (2002) document the debate over the efficiency of the residential real estate brokerage (RREB) industry. They note that

> if the RREB industry is deemed grossly inefficient, the implication is that over time, with new innovations, commission rates (service prices) would come down and/or services would increase or improve. On the side arguing for general efficiency are Lewis and Anderson (1999) and Anderson, Lewis, and Zumpano (1999). On the other side are Miller and Shedd (1979), Crockett (1982), Wachter (1987), Yinger (1981), and others. A key premise behind those arguing for inefficiency is the fairly uniform and rigid commission pricing within local markets for similar property types. Prices appear to be abnormally stable for a competitive market. A rare exception to this widely held belief was presented by Carney (1982).

Delcoure and Miller (2002) provide a comparison table (Table 6.10 as a shorter version) of brokerage costs across a cross-section of countries. Globally, they find, compared to the US, much lower residential commission rates in most of the other highly industrialised nations, including the United Kingdom (UK), Hong Kong, Ireland, Singapore, Australia and New Zealand (see Table 6.10: International Commission Rate Comparisons). Brokerage fees in Hong Kong, which is typically about 1% for the seller, are among the lowest in the world. In the UK, the commission rates average less than 2%. In New Zealand, South Africa and Singapore, commission rates average more than 3%. Many countries have 5% or less fees, including Germany, Spain, Israel and Thailand. Indonesia, Jamaica, Sweden, Trinidad and Tobago, and the Philippines also have about 5% fees. The authors note that "it is hard to argue that non-US countries have more efficient communication technology, real estate public information or record access that would lead to lower commission rates." Commission rates in less developed countries, such as Russia and Belarus, are not very reliable as there are less public records and no reliable MLS (Multiple Listing Services), and the fees can range between 5% and 15%. In China, there are regulatory environment and license standards for its real estate industry and a very high transfer tax of 15%.

Research issues related to brokerage

The market for real estate is huge with approximately 81% of single-family dwelling sales and total brokerage commission fees adding up to more than $65.5 billion a year (Rutherford et al., 2005). Academic research has concentrated on several aspects of the brokerage industry. Two roles of agents have been examined: matching and bargaining. Often, various standard economic theories are used such as:

- price theory to contract theory;
- matching games to bargaining games;

Table 6.10 International real estate brokerage commission rate comparisons

Australia	5% on the first $18,000, 2.5% -thereafter; also properties are sold through auction system; advertising is provided by real estate agent.
Brazil	5% commission, less on a higher priced units.
Canada	3-6% commission rate. An agent handles on average 3 to 5 sales per year.
China and Hong Kong	No set regulations and standards for a real estate transaction in China. Commission fees vary from 5% to 10%. Also, there is a 15% real estate transfer tax. However, Hong Kong has a significantly lower real estate brokerage fee of typically 1% for the seller. Hong Kong does not require dual representation and one agent may deal with both the buyer and seller. However, both parties typically have separate lawyer representation. In Hong Kong, the maximum transfer tax is 3.75%.
France	Yes Only 50% of property sold is listed with a real estate agent; real estate transactions are kept very private; 50% of the real estate is sold by owner.
Germany	Negotiable commission rate that varies from 3% to 6%.
Greece	4% commission rate, where the buyer and seller are responsible for 2% each. Also, there is a 12% value added tax on a real estate transaction.
Indonesia	5% paid by either buyer or seller, but not both; a buyer's broker is required for real estate transactions.
Japan	3% commission rate.
Malaysia	3% on the first $100,000, and then 2% of the remaining amount of the sale; commission is paid either by buyer or seller, not both.
Mexico	Varies 5-10% commission rate. Large emphasis on MLS.
Netherlands	Yes 1.5-2%, broker represents either the buyer or the seller but not both. The seller pays the fees.
Russia	5% to 10%, but 'net listings' are common; advertising is provided by real estate broker/agent; FSBO very common; buyer broker representation is not required. Some commissions are set in dollar or ruble fee amounts. Reliable market information is difficult to acquire.
Spain	Commission rate depends on the property location, averaging 5% of total estate price.
Sweden	5%; commission is paid by seller. 10% commission is typically charged for lower priced units.
United Kingdom	1%-2% is typical; in very competitive areas 0.5-0.75%; in low priced areas as high as 3.5%. Advertising is provided by real estate broker/agent; buyer broker representation is not required.
United States	6%-7%; advertising is provided by real estate broker/agent. In 1999, some real estate agents charged flat fees that ran from 2 to 4%. Auctions are increasing but usually at the same fees or higher than normally charged by the brokerage firm involved.

Source: Based on information from Delcoure and Miller (2002); for a full list of the countries and analysis, refer to the source. http://www.umac.mo/fba/irer/papers/past/vol5_pdf/012_039US.pdf

- market structure to the theory of the firm;
- moral hazard to adverse selection problems;
- agency theory to search theory.

It is often argued that there is a 'Principal-Agent' issue, i.e. when an owner appoints an agent to sell a property, he/she may have less/low quality information than the agent and be disadvantaged in setting reservation price and negotiating. So, the question becomes: what is the benefit of going through an 'agent'?

- *What are the gains in terms of 'time-on-market' versus 'price uplift'?*

- When an agent sells or buys his/her own home, how does it 'differ' from his/her clients' transactions?

It is argued in the literature that real estate agents should act in the best interest of the clients and act similarly to what they would do if it were their property up for transaction. This should be a part of their fiduciary duties. However, given the incentive structure and industry practices, it is evident that the interests of the agent and the seller may not converge. This is due to the fact that the standard commission structure gives agents only a fraction of the price, about 5–6%, which is much smaller than the seller's stake and it thus creates positive externalities. The seller's goal is to maximise price uplift and minimise the time on the market and the agent's goal is to maximise net commission revenue and minimise marketing time and effort. Therefore, the economic interests are not aligned and are in conflict with several implications for the transaction and market (see Anglin and Arnott, 1991; Geltner et al., 1991; and Miceli, 1991). It can also be argued that the agent's motivation may as well be directly opposite to the price maximisation strategy of the seller to achieve faster sales. Rutherford, Springer and Yavas (2005), analysing US data, find that agent-owned houses sell no faster than client-owned houses, but they do sell at a price premium of approximately 4.5%. In a later study, Rutherford, Springer and Yavas (2007) also find that real estate agents receive a premium of 3.0–7.0% when selling their own condominiums in comparison to similar client-owned condominiums. This implies that agency problems can exist even in more homogenous condominium markets, compared with the more heterogeneous single-family home markets. The agency problem may be less severe for condominium sales due to longer time on the market for agent-owned condominium units. It is, however, also argued that competition among agents may completely eliminate the agency problems created by the percentage commission structure. Williams (1998) and Fisher and Yavas (2010) have developed theoretical models that show such possibilities.

The role of an industry association has also been explored. Huang and Rutherford (2007) explores whether or not the REALTOR® designation (National Association of REALTORS in the US) serves as a signal of the effectiveness of an agent. They compare the price and time-on-the-market for realtor listings with those of non-realtor listings on the multiple listing service (MLS). The authors find that non-realtor properties list and sell at a lower price, take slightly longer to sell and are less likely to sell than properties listed by REALTORs in an MLS setting. So, the question then can be posed: how do we evaluate the 'quality' of brokerage services, given it is mostly about non-price competition. Shall we set a minimum standard for the level of services? If we do so, what happens to the competition in the market?

Nanda and Pancak (2010) provide analysis on such laws. Ten states in the US have enacted laws requiring a real estate broker to provide a minimum level of services as noted previously. The US Department of Justice (DOJ) Antitrust Division and the US Federal Trade Commission (FTC) both considered these types of requirements as anticompetitive, and both agencies argued against state enactment. As the authors note, "these types of laws are deemed anticompetitive, primarily because they prevent a limited-service real estate broker from contracting with a seller to provide only access to the brokerage multiple listing services (MLSs) for a flat fee." They presented evidence that indicates that state REALTOR associations have been primary supporters of state minimum service laws. Support for minimum service laws is clear, that real estate consumers should expect certain services from a broker and laws can guarantee a minimum level of services (DOJ and FTC, 2007). In addition, if a seller does not get desirable assistance from the broker, he/she can ask for assistance from the buyer's broker. It is also claimed that the buyer's broker needs protection for assisting a seller because the seller does not pay the buyer's broker. Moreover, helping the seller may also create a dual agency conflict of interest.

The state agencies' efforts have experienced mixed reactions. Alabama, Idaho, Missouri and Texas passed minimum service laws regardless of the federal opposition. New Mexico, Tennessee and Michigan, however, changed proposed legislation to make minimum service laws waivable or optional. As a result, there were strict and light versions of the law. Illinois became the first state in 2004 to adopt minimum service requirements. The legislation was constructed with the following statements (Nanda and Pancak, 2010):

- "Accept delivery of and present to the client offers and counteroffers to buy, sell, or lease the client's property or the property the client seeks to purchase or lease.
- Help the client develop, communicate, negotiate, and present offers, counteroffers, and notices that relate to the offers and counteroffers until a lease or purchase agreement is signed and all contingencies are satisfied or waived.
- Answer the client's questions relating to the offers, counteroffers, notices, and contingencies."

The Illinois Association of REALTORS supported by arguing that these laws can promote greater professionalism and accountability within the industry. In an academic study, Miceli, Pancak and Sirmans (2007) argued that minimum service laws may support a brokerage compensation scheme that may not necessarily serve the consumer's interest. Similarly, Magura (2007) put forward a critique that minimum service laws may entail a chilling effect, due to broker price-cutting actions. White (2006) also was of the opinion that mandatory minimum service requirements for sellers' brokers might cut the competition from discount brokers. Levitt and Syverson (2008) questioned if there were any justification with respect to consumer protection for these laws. They found that houses sold by limited-service brokers took longer to sell but there were no price uplifts. Nanda and Pancak (2010) find that factors reflecting state brokerage influence, e.g. strong industry associations and broker membership on licensing boards – are not the main drivers behind the enactment of minimum service laws. Factors related to consumer protection motivations such as greater number of complaints against brokers, stricter prelicensing requirements, and a Democratic state legislature can increase the likelihood of law adoption. This leads to another pertinent question regarding the effectiveness of such regulations in the industry.

A number of studies has looked at inefficient overproduction of real estate brokerage services (Yinger, 1981; Miceli, 1992; Yavas, 1992;, Turnbull, 1996). In a more recent study, Hsieh and Moretti (2003) analyse the productivity of real estate brokers (measured by houses sold per hour worked) and their findings indicate a statistically significant drop in productivity if the land price goes up. A possible reason, they put forward, could be overproduction of services (i.e. unnecessary increase selling costs), driven by fixed commission structure and entry of new brokers. Similarly, studying the costs of brokerage services, Han and Hong (2011) find that a 10% increase in the number of brokers will increase costs by almost 12.4%. The authors further attributed more than one-third of this effect to wasteful non-price competition. Nanda et al. (2016) examined whether the framework of laws and regulations that govern real estate brokerage industry can reduce costs due to creation of an environment supportive of more efficient unbundling of brokerage services (i.e. offering various service-by service fee structure). They argue that, if it does reduce the costs, then the number of full-service brokers would correlate with house price changes, transactions and law change. Ease of entry and fixed nature of commission rates are argued to be the main factors driving inefficiency. They find that the enactment of waivable minimum service and non-agency provisions reduced growth in full-service brokers, i.e. laws allowing transactions facilitators or limited-service brokers tended to clarify brokerage duties, reducing legal risk and encouraging unbundling of brokerage services. However, results are not robust to unobserved heterogeneities related to specific time periods. As the authors point out, this is likely because of the clustering of regulatory changes in the middle of the 2000s. Beck, Scott and Yelowitz (2012) is of the opinion that the natural monopoly aspect of the multiple listing service (MLS) can facilitate these high commission fee structures. MLS can allow brokers on both sides to share information in a way that can enable them to cooperate to complete a transaction, which gives them significant ability to charge high commission rates. Yavas and Colwell (1999) and Woodall and Brobeck (2006) also identified MLS as the major reason for high commission fees and too many brokers in the market.

Nanda et al. (2016) further note that improvements in technology, innovations in technology platforms and recognition of the value it can generate for the investors can yield significant benefits. This can be achieved by reducing human interference and time cost and moreover, information flow with easy access to information and fast decision-making by prospective homeowners can be facilitated by technology interventions, which I discuss in more detail in the last two chapters. Some brokers or estate agents have started to provide unique selling propositions. This can substantially add value to the services and can facilitate unbundling of the traditional full package of brokerage services. Purplebricks.com is a good example. Technology allows identification of possible houses to purchase and this can be undertaken by online platforms with property details. Viewings could be undertaken by virtual tours. Price information is available through online portals. Local area information can also be found from online portals and Google satellite and associated land-based information on the urban form. Therefore, the matching role of the real estate brokerage can be greatly diminished by web-based technology. So the relevant questions are:

- Does that mean bargaining role will be the sole area of intermediation, as matching will be facilitated by technology?
- In that case, would it lead to more intense non-price competition among agents? If so, what would determine the service quality?
- Would it create more entry barrier as experience, knowledge and economies of scale play bigger roles?
- How does it change the housing search patterns and behaviour?
- What are the implications for estate agents' future role?

In this chapter, I have discussed the housing search issues, tenure choice and aspects of housing mobility. Much of these crucially depend on and are driven by demographic patterns of householders. Tenure choice for natives and non-natives differ significantly. The same can also hold for young and old people. In the next chapter, I discuss effects and implications of demographic changes such as the ageing population and migration in the housing market.

Selected research topics

- Deposit constraints, ideology of homeownership – policies that have led to polarisation of tenures.
- Case for public housing and role for private developers.
- The user cost of housing as a framework for identifying any bubble.
- Role of online technology in real estate transactions.

7 Demographic shifts and housing

Chapter outline

- Migration and housing
- Housing for young people
- Population ageing and housing
- Policy issues for retirement housing provision
- Selected research topics

In this chapter, three major cohorts of the demographic profiles having significant implications for the housing markets – migrants, young and elderly people – are discussed.

Migration and housing

Arguably, the most consistent and persistent socio-economic trend is migration. It is a long-run mega-trend. Migration can be defined as movement of people from one spatial unit to another spatial unit. We observe different types of migration at different times across different regions. Some migration is international (people moving from one country to another country), some are inter-regional (moving from one region to another within a country) and some are intra-regional (one part of the region or city to another). Housing issues are very much linked with all types of migration – with much implications for demand for housing. Sometimes migration is permanent or long-term, having significant and persistent impact on the housing market; other times it is transient, still having significant impacts in the housing market.

Throughout the human history, migration has led to exploration of new places. People have been seeking better opportunities, pursuing ambitions of better quality of life; and leaving areas of conflict, war, famine and risky environments. More recently, the intensity and spread of migration has affected several major regions of the world. Due to dynamics of economic geography, societal composition and structure have become more complex, complicated and challenging (Spencer, 2002). Migration can have significant effects on a local community in several ways. A persistent migration over a number of years can also change the make-up of the local community. For example, since World War II, migration is a global concern with much transformation in the nature of emigration and immigration (Becker and Morrison, 1999).

We can possibly trace the earliest human migration back to several thousand years ago when a group of ancient humans from Africa embarked on journeys to distant parts of the planet and started settling. This early migration could have been driven by major climatic shifts, which could possibly cause extinction of the human race. With climatic conditions becoming more conducive for human living, human population flourished and members of human society started journeys that took them far and beyond Africa to the coast of India, Southeast Asia and Australia. Soon another group embarked on an inland trek to the Middle East and Southern Central Asia. After a few thousands of years, a small group of the Asian hunters and gatherers migrated to the East Asian Arctic. Large ice sheets covering the far north formed a land bridge that connected Asia to the Americas and within another thousand years humans settled in South America. The strength of human population was not great during those early years. However, improving understanding of agriculture and an abundance of food supply due to agricultural production in the recent millennium helped with the expansion of population. The National Geographic constructed an important map with the earliest human journey or migration in search of better climate, better food and other resources. According to the study, humans first ventured out of Africa some 60,000 years ago. The National Geographic's Genographic project mapped the appearance and frequency of genetic markers of modern humans to document when and where ancient humans moved around the world. These great migration decisions have shaped the agglomeration of human civilisation that we see today. Africa is where human species went through early evolution with the earliest fossil record found at Omo Kibish in Ethiopia, around 200,000 years ago (see National Geographic Genographic website for details). As per the project findings based on the genetic and paleontological record, human species only started to migrate out of Africa between 60,000 and 70,000 years ago. The most likely reason is perhaps major climatic changes at the beginning of the last Ice Age. With those climatic shifts, food sources might have become scarce along with an inhospitable cold environment and lack of habitable shelter. The findings from the project also suggest that the earliest humans might have colonised the Eurasian landmass by crossing the Bab-al-Mandab Strait separating present-day Yemen from Djibouti. These early migrating humans continued further along the coast to India and other parts of Southeast Asia and reached Australia by about 50,000 years ago.[1]

With discoveries and sharing of information and knowledge exchange along with documenting of habitable land masses, migration has become intense. Due to voyages undertaken by our pioneering explorers over the last 500 years, migration became a widespread phenomenon with much impacts across the world. The migration trends in the last century has been quite intense with rapid displacement and settlements. Globalisation and easier access to the global market coupled with information flow and sharing shaped much of the recent migration events (Massey, 2004). With the renaissance and success of Western civilisation, the last 5–6 decades have experienced almost unidirectional migration, i.e. towards Europe and North America. For example, several major migration events occurred in the last century – British migration to Australia, New Zealand and South Africa after World War I; Turkish migration to Germany after World War II; Indian sub-continent to the UK after the Indian independence; Ugandan Asians to the UK in the mid-1970s and persistent migration from South America to the US since the 1950s. Sometimes it is the active government policies that have attracted economic migrants. At the same time, policies have been frequently devised to curb migration. Often, migration issues take significant political importance, resulting in divisive, polarising views and policy outcomes. Figure 7.1 shows the total migrant stock from 1990–2017 for a selection of countries.

1 Source: https://genographic.nationalgeographic.com/human-journey

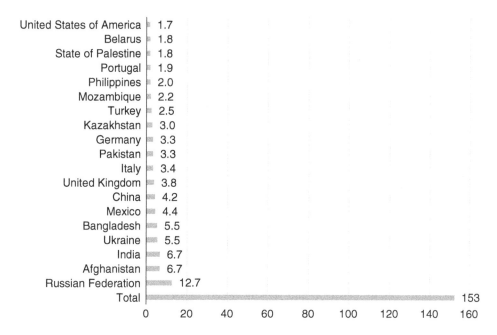

Figure 7.1 Total migrant stock (in millions) at mid-year by origin and by major area, region, country or area of destination, 1990–2017.

Source: Based on data from United Nations, Population Division; for a full list, refer to the *source*. http://www.un.org/en/development/desa/population/migration/data/estimates2/estimates17.shtml

Several reasons drive the decision for individuals to move – economic, social, political and environmental. Economic migration is linked to movements undertaken to find employment or to pursue better prospects of income or to follow a professional career. Educational attainment is often linked to economic migration. Familial links are associated with social migration, which can also be triggered by aspirations of having better quality of life. However, political and environmental migration are linked with a significant push factor such as political turmoil, civil war or other types of war, natural disaster, environmental disasters, etc. Throughout human history, we have experienced all these types of migration, which have contributed to economic development of some regions but also led to economic declines of some regions.

Migration has huge impacts in shaping places, societies, markets and outcomes – with favourable and adverse implications. Both the places that migrants move out of and places migrants move into are hugely affected by the movements. Internationally, emigration (i.e. individuals leaving a country) and immigration (i.e. individuals entering a country) have been greatly studied in the academic literature and greatly debated in the political sphere. Hollifield (2004) pointed out that migration can be broadly driven by both pull and push factors. Pull factors can be based on attraction, encouragement or motivation, while push factors can be based on frustration, disappointment and threat to life and property.

Some of the **common push factors** are: lack of security, safety issues, agricultural declines, lack of food and other resources, natural disasters (like floods, droughts, earthquakes, etc.), persistent poverty and all types of wars. Natural causes have been very common in the earlier periods of civilisation, while conflict-related displacement of people have been more common in recent millennia. Some **common pull factors** are: better employment prospects, better income levels, better agricultural yields, quality of life, safe and secure living from crime and natural disasters, favourable political conditions, more conducive environment, good climatic conditions, etc. Within all types of migration there are usually a complex combination of numbers of push factors and pull factors. These factors have been major areas of research. Some of the pull and push factors can be just the opposite of each other, e.g. employment prospect/ high unemployment rate. Shaw (1975), however, observes that push and pull forces may not necessarily manifest equally and should therefore be analysed in the context of demographic, socio-economic and life cycle factors in order to have a deeper insight into the phenomenon.

Developing and developed countries differ significantly in terms of migration patterns and implications. First, increasing levels of emigration in developing countries have significant implications for the countries losing talents. This on the other hand has led to an increased level of immigration across developed countries. Several authors have documented and commented on these patterns (see Kingdon et al., 2006; Duru and Trenz, 2016; Gibson et al., 2017).

Immigration has complex and widespread impact on the countries receiving the population. Although mostly positive through an increase in labour force availability, it can have adverse implications as well. Countries like the United States of America has greatly benefitted from waves of immigration and subsequent bursts of innovation and productivity increases from the immigrant population. Migration is a complex and global issue. As it can be variously associated with a number of factors such as economic, social, environmental and political, it is a very policy-relevant subject area. With widespread global migration, it is increasingly difficult for national governments to manage its effects so that positive effects and net gains for the country and society can be ensured.

Immigration has led to significant population growth, particularly in developed countries. OECD 2017 reports that immigration is at its highest level since 2007. For example, over 50% of the population growth in the UK for the next two decades may be associated with net migration (ONS, 2014; Migration Directory, 2016). A staggering number of people are moving – almost 3% of the world's population, i.e. approximately 192 million people are estimated to be living outside their place of birth (Organisation for International Migration, 2017). Over the years, migration has taken many complex features and the changes are fast, although tracking migration has become more accurate due to technology interventions at borders and identification systems. Organization for Economic Cooperation and Development (OECD) countries are by far the largest destination of immigrants (OECD, 2017). Along with the United States of America, the United Kingdom is a global destination and one of the countries with the highest net migration rates in the world. Due to historical and colonial links with several parts of the world, the UK has experienced large-scale migration from many parts of the world. At the same time, many Britons have also emigrated to several countries. According to OECD (2017), the population of foreign citizens in the UK went up to 5.85 million in 2016, a 6.4% increase from the previous year. With the UK population reaching almost 74 million by 2039, net migration will be a major contributor having almost 50% share of the population growth of the UK over the next 25 years (ONS 2015). New member states being added to the EU since 2004 have resulted in the addition of much more EU citizens in the UK compared to non-EU citizens. Another reason is the tightening of legal requirements for entry of non-EU citizens.

Immigration in general is seen with much scepticism in recent years. With the rise of populist politics, several countries have seen intense debate on migration. There are economic reasons for such outcry. With population increase comes increasing demand for public services and state-provided services such as social care, healthcare, affordable housing, etc. Increasing demand leads to more competition. Along with the 2008 Global Financial Crisis, which has led to worsening economic and financial conditions across many developed countries, increasing demand for public services and the rising fear of worse quality of life have led to gains in popularity of nationalist sentiments. Recent years have experienced much success of such nationalist movements. Political outfits supporting nationalist sentiments have flourished across the developed nations. For example, the United States has experienced polarisation of national politics with power being vested with the populist protagonists. The United Kingdom has seen a referendum which resulted in the UK poised to exit the EU in March 2019 (the so-called 'Brexit'). Mismanagement of immigration and lack of effective policies to support both native and non-native population, along with harsh economic realities, can lead to the emergence of nationalist sentiments. Regardless, as history has repeatedly shown, human migration will continue with bumpy rides and the eventual arrival at the destination, which is so often the story of an individual migrant.

Immigration helps in many ways with positive economic outcomes. However, Spencer (2003) notes that most countries are still not adequately prepared to manage immigration and part of the reason is much faster speed and depth coupled with political anxieties related to governance arrangements and difficulty in dealing with potentially costly political trade-offs. Effective management of fast and complex migration requires significant investment in technology, staff recruitment and robust regulatory frameworks, much of which depends on taxpayers' money, which in turn requires strong political support.

Scholarly works on migration have taken multi-disciplinary approaches. Given the complex nature of the subject, several disciplines need to come together to analyse the issues, challenges and implications. The processes, issues and challenges have been commented by several authors. Typically, the increase in immigration may lead to changes in the life-courses and education, labour market and housing outcomes and pathways of natives and non-natives. Changes observed in the outcomes and pathways are often associated with socio-economic factors, demographic factors, household composition and sociocultural factors as well as other unique location-specific factors.

Housing is a very important consideration for migrants and all types of migration involve housing outcomes with complex issues and dynamics. There are significant differences in housing pathways for natives (born in the country to parents who are also born in the country) and non-natives (first generation migrants – born outside the country, and second generation migrants – born in the country to migrant parents). Housing issues of the migrant population have been studied extensively by the academic community. Much of those works informed the policymaking mechanisms. With housing being a basic need in terms of shelter, the consumption motive as well as the most important wealth-creating channel for natives and non-natives, it is very widely studied. Due to its complexity, it is important that causes of migration are studied along with the outcomes.

Migration issues across OECD countries have been frequently studied by the researchers. All studies document significant changes to the local housing market due to migration. However, whether immigration leads to positive or negative effects is still debatable. Housing pathways have been studied in detail with a focus on factors, mechanisms of accessing housing finance, any biases that are not explained by economic factors (i.e. discriminations) and the outcomes.

Some research have studied housing issues in conjunction with labour market issues, as for the migrant population, employment is often the first priority. Tenure choices at various stages of the life cycle (upon arrival) have been studied. While renting is an obvious choice upon arrival, much attention has been paid on understanding timing and mechanism of changes in tenure, e.g. private renting to homeownership. An established scholarly body of work relates to housing tenure choices for ethnic groups (see Skifter Andersen et al., 2016). These studies associate the tenure outcomes to a households' aspiration, information set, future expectation and the life cycle. Several national levels of demographic data sources indicate consistent and significant differences in housing tenure choices of natives and non-natives, e.g. data from the US Census Bureau (2011) in Trevelyan et al. (2013), as well as ONS Labour Force Survey (2016) and Migration Observatory (2016). The Migration Observatory (2016) report finds that the prospect of homeownership is likely to increase as immigrants have spent more time in the country with decreasing reliance on renting options. It is important to note that this may not necessarily be the case always, as other factors may explain systematic variation in the outcomes (see Coulson, 1999). Individual, household and demographic factors may influence the migration effects. Cultural differences in perception towards housing and savings behaviour can effectively explain how soon or late the tenure choices are changed. It is important to bring in the lessons and insights from the life cycle theory of Modigliani and Brumberg (1954). As per the life cycle theory, much of the enablers of housing outcome such as income, education and savings can be determined by the age. For the migrant population, time in the destination country can prove to be a better determinant of the outcome as several financial and economic criteria get activated upon entry which can determine homeownership decision points.

Since the early 1990s, due to technological progress and the talent demand from several developed countries, global migration flows have become larger in scale and complex with more heterogeneity in types and patterns. In several countries, including in the UK, the detrimental effect of immigration on the wellbeing of the settled residents has been often put forward with much political content in it. However, empirical evidence is thin. The local market effect is not well-understood and documented. Robinson (2010) tried to fill this gap in understanding by outlining a framework to support the exploration of neighbourhood effects of new immigration with three types of explanations for geographical variations in local experiences of new immigration: the individuals living in a place; the opportunity structures apparent in the local environment and the sociocultural features of local communities. As the author aptly notes,

> Place is a social and material setting and 'meaningful location' (Agnew, 1987). It represents the context within which new immigrants and settled residents come together and possesses the potential to inform variations in both the impacts of new immigration and how people and institutions make sense of these consequences for the local area and their own well-being. This is not to deny that the differential package of rights and opportunities associated with different immigration pathways represents an important determinant of the arrival experiences of new immigrants. Nor is to neglect the fact that new immigrants can prove resourceful and exercise agency even within the most constrained of circumstances. The point is that the experiences and impacts of new immigration cannot be fully appreciated without the application of a geographical perspective on place.

The debate around the impact of immigration on the local housing market is quite intense in America due to a strong immigration flow and the historical importance of immigration in the country's prosperity. Saiz (2007) studied if there is a local economic impact of immigration. Immigration tends to put upward pressure on rents and also subsequently housing values in US destination cities. Rent growth and immigration have often been found to have positive correlation in historical data. Saiz (2007) used instrumental variables based on a 'shift-share' of national levels of immigration into metropolitan areas. Quite interestingly, he found that immigration inflow equal to 1% of a city's population is associated with almost 1% increases in average rents and housing values. This is a very significant result with a high magnitude of the effect. As Saiz (2007) rightly asked: "but why should we be specifically interested in the impact of immigration on rents? How is immigration different from general population growth? Is it surprising to find a substantial impact of immigration on local housing markets?"

Immigrants are different from the general population. Especially with regards to housing market outcomes, there are many distinguishing factors. Saiz (2007) notes that immigrants are much more spatially concentrated than natives. For example, in 20 major metropolitan areas, from 1983–1997, more than half of the new immigrants settled in only ten MSAs, which contained just 20% of the metropolitan population. Compared to about 20% of Americans who lived in non-metropolitan areas in 1980, only 4.34% of immigrants admitted during the 1983–1997 period settled outside metropolitan areas. This implies that the effect of immigration can be expected to be stronger on specific housing markets. However, it is not very clear what would be the impact of immigrants on the housing markets of destination cities. Muller and Espenhade (1985) documented that the rental housing market had experienced major upward pressure during 1967–1983 in terms of prices in Los Angeles compared to other US metro areas. Los Angeles has been one of the most important 'gateway' cities for immigration, especially in that period. There were documented evidence from other countries with significant immigrant population and popular destination. For example, Burnley et al. (1997) found significant association between immigration and short- and long-term inflation of housing prices in Sydney. Sydney is an important 'gateway' city in terms of immigration in Australia. For another important immigrant destination country, Canada, Ley and Tuchener (1999) also found noticeable correlation between house prices and immigration in Toronto and Vancouver, two main 'gateway' cities in Canada. However, it is important to note, as per Saiz (2007), these studies do not control for confounding factors, other variables that could explain the variation. Without effective controls for other variables, it is not possible to statistically attribute the variation to immigration only. The descriptive nature of these studies should therefore be considered with caution. However, there are still important findings that should be investigated with appropriate methodological rigour and a strong information base.

In terms of econometric analysis, it is challenging to separate out the association between immigration and house prices or rents. Two major problems that can severely bias the estimates and lead to misplaced inferences – unobserved heterogeneity and endogenous feedback loops. As Saiz (2007) aptly noted:

> firstly, omitted variables that are not observed by the researcher could be driving both immigration inflows and housing costs. Immigrants may respond to other factors that cause rents to increase, such as expectations of future economic growth, improved amenities, or changes in the preferences for existing amenities. In principle, this could lead the researcher to overestimate the impact of immigration on rents. Secondly, immigration could be endogenous. Immigrants may be looking for better deals: they might settle in areas where rents are increasing more slowly. If immigration inflows are very sensitive to housing costs, then the estimates of the relation between immigration and rents could

be biased downward. In this context, one needs to look for exogenous sources of variation in the immigration inflows to ascertain causality.

In an earlier paper, Saiz (2003) provides evidence of a causal relationship between immigration inflows and housing rents. The author used the "Mariel Boatlift" as a natural experiment, following Card's (1990) study about labour market outcomes. The Mariel boatlift was a mass emigration of Cubans from Cuba's Mariel Harbour for the United States between April 15 and October 31, 1980. The refugees were called 'Marielito' (or, plural 'Marielitos'). The exodus might have been driven by a sharp downturn in the Cuban economy, which have also resulted in generations of Cubans immigrating to the US for better economic outcomes. On April 20, 1980, the Castro regime announced that people, at their will, could emigrate to the US by boarding boats at the port of Mariel, west of Havana, launching the great migration event called the Mariel Boatlift. The boatlift was driven by housing and job shortages which caused internal tensions in Cuba. The exodus was finally ended by mutual agreement between the US and Cuban governments in October 1980. Overall, 125,000 Cubans reached US shores in about 1,700 boats. Some migrants died due to boat capsizes[2]. This immigration shock represented an exogenous increase of 9% in Miami's renter population in a single year, which was equivalent to about 4% of the total population. Saiz (2003) finds that rents increased from 8% to 11% more in Miami than in the comparison groups between 1979 and 1981. By 1983, the rent differential was still at 7%. Rental units of higher quality were not affected by the immigration shock. Units occupied by low-income Hispanic residents in 1979 experienced an extra 8% differential hike with respect to other low-income units. Relative housing prices moved in the opposite direction from rents in the short run. Although immigration is perhaps a very likely explanation for this differential growth in rents, there are potential biases in the differences-in-differences approach that Saiz (2003) used. As Saiz (2007) points out

> a similar argument applies to competition in the housing market. Immigrants may be less sensitive to housing rents, because local immigrant-specific amenities and networks are more important for them. Natives, though, may be more sensitive to local rents. If this is the case, immigration inflows could spur net out-migration of natives because of the increased housing costs that are associated with a housing demand shock. There is no way to separate the effect of increased housing demand (immigration) from the potential decreased demand associated with potential native out-migration. Part of the local response to the treatment (immigration) can occur through native out-migration. In this case, we need to be careful about the interpretation of the coefficient of immigration on rents. In general it will not correspond to the housing supply elasticity. Nevertheless, we should expect a positive effect of immigration on rents if natives are not extremely sensitive to changes in housing costs, and if they are not displaced "one-for-one" in the labour market.

Saiz and Wachter (2011) examined potential unobservable shocks that may be correlated with proximity to immigrant enclaves. Using a geographic diffusion model to instrument for the growth of immigrant density in a neighbourhood, they investigated if within metropolitan areas, neighbourhoods of growing immigrant settlement are becoming relatively less

---

2 https://www.history.com/this-day-in-history/castro-announces-mariel-boatlift

desirable to natives. The evidence is consistent with a causal interpretation of an impact from growing immigrant density to native flight and relatively slower housing value appreciation. Further evidence indicates that these results are driven more by the demand for residential segregation based on ethnicity and education than by foreignness per se.

In a more recent work, Accetturo et al. (2014) examined the impact of immigration on segregation patterns and housing prices in urban areas. They developed a spatial equilibrium model that allows them to investigate how the effect of an immigrant inflow in a district affects local housing prices. This is revealed through changes in how natives perceive the quality of their local amenities and influences on their mobility. The model formalises how an immigrant shock in a district affects local amenities, thus influencing local prices and mobility aspects of the native population. The model is developed with several assumptions. One of the crucial assumptions is that the natives' indirect utilities would depend on both real wages (and, therefore, on housing prices) and the perceived quality of amenities. Interestingly, the perceived quality of amenities is assumed to be a function of the number of immigrants residing in the district. While the authors have not imposed any restrictions on the relationship between immigration and natives' perception of local amenities, natives are allowed to have racial or religious preferences and that could induce a negative effect. This negative effect may stem from communication challenges, deterioration of local standards of living due to the crowding effect on local indivisible public goods (e.g. public parks, public transport, publicly-funded healthcare facilities), etc. On the positive side, natives may have preferences for cultural diversity and would be inclined to enjoy the benefits of more variety in the local goods (e.g. ethnic restaurants). These two effects would be at work with uncertainties for the net effects. The authors posit several theoretical predictions: (i) an immigrant shock to a district may increase the average price of housing at the city level; (ii) the districts facing the immigration shock will have higher (lower) growth in house prices than the city average if and only if migrants have a positive (negative) effect on the natives' perception of local amenities; (iii) an immigrant shock may displace natives who may prefer to move to other areas of the city; native mobility is affected by an income effect (i.e. the crowding out of natives due to the increased demand for housing by immigrants) and an amenities effect, whose sign depends on the effect of immigrants on local amenities; (iv) if immigrants are concentrated in a district with a more rigid housing supply, there may not be any additional effects on prices but a stronger native outflow may be observed. Predictions of the model are tested by using a novel dataset on housing prices and population variables at the district level for a sample of 20 large Italian cities. To address endogeneity problems, the authors have adopted an instrumental variable strategy which uses historical enclaves of immigrants across districts to predict current settlements. With these controls and the model set-up, they find that immigration raises average house prices at the city level, while it reduces price growth in the district affected by the inflow vis-à-vis the rest of the city. This is an interesting finding. It implies that the pattern may be driven by native-flight from the districts with a heavy concentration of immigrants towards other areas of the same city and this indicates a negative effect of immigrants on native's perceived quality of local amenities.

In another recent paper, Mussa et al. (2017) analysed the effect of immigration on the US housing market, both in terms of rents and single family house prices. Similar to Accetturo et al. (2014), they also model the housing market in a spatial econometrics context using the spatial Durbin model. There are significant benefits from using a spatial Durbin model. It can help in examining and estimating both the direct and indirect effects of immigration inflows on the US housing market. They find that an increase in immigration inflows into a particular MSA is associated with increases in rents and with house prices in that MSA. More interestingly, they also find that the increases in one MSA seem to drive up rents and prices in neighbouring MSAs. These results are consistent with native-flight from areas of heavy concentration of immigrants and popular destination for the new immigrants due to socio-economic network effects. As the author aptly notes,

> the timing of immigration (immigration vintage effects) may also influence relative housing values. For example, Akbari and Aydede (2012) note that immigrants tend to buy housing after a period of residency in their adopted homelands, while recent immigrants tend to rent. We might, therefore observe excess demand for rental housing units driving up rental prices when immigrants first move in. But over time, as immigrants move onto becoming homeowners, they may apply upward pressure on prices for single family homes. Hence the vintage of immigrants may affect different segments of the housing market in different ways. Using the Mariel boatlift natural experiment, Saiz (2003) concludes that this population influx raised rents and depressed home values in immigrant receiving areas. The Mariel immigrants tended to rent, possibly propelling natives to purchase housing in other areas. Using finer distinctions (census tracts instead of MSAs) Saiz and Wachter (2011) conclude that housing appreciation was slower in immigrant neighbourhoods, with natives moving out in response to the in-migration of the newcomers.

Literature so far has reported mixed views and conflicting evidence on the relationships among immigration, housing tenure, spatial choices and housing demand. There is no doubt that there are a number of complex issues on this topic. Several theoretical predictions, when tested empirically, face challenges in terms of real world relevance and applicability. Therefore, the researchers have had divided opinions on the detailed mechanisms of the issues. Housing demand is a complex subject and is a reflection of a complex combination of several observed and unobserved factors. Migration-related housing matters take on additional complexities due to issues related to settlement and the labour market. The interdependencies in non-natives' housing outcomes can be a lot more complex compared with the natives. Often several decisions are jointly made which makes identification and separation of the issues particularly challenging and often lead to misplaced inferences. Goodman (1990) notes that demand for housing rental and homeownership are jointly related. Although, Goodman (1990) provides an account of the variations of ethnic and racial demand in the context of the life cycle and future expectations, as well as demographic, individual, household and housing characteristics, it does not provide significant insights on the impact of immigration. Skifter Andersen et al., (2016) noted that the migration cycle may influence housing tenure preferences, leading to further impact on housing demand. It can be generally expected that the changes in tenure for migrants are more predictable. Shifting from renting to owning is often seen in the market for the immigrant population. This shift is very much an important part of the maturity in settlement and integration to the destination country. This shift though creates much stronger demand-related forces in the market, e.g. demand for mortgage products and type of housing.

Beyond the impacts already discussed above, immigration can also lead to a few other impacts. It can lead to demand shocks due to the quick, large influx of immigrants into an area that the local market is not used to experiencing. Such shocks can lead to significant disruption in the 'normal' demand-supply interactions in the short-run. If not checked and acted upon with supply-side measures, this can lead to significant price fluctuations. For example, Skifter Andersen et al. (2012) found that surplus demand in Copenhagen had led to a rent ceiling which might have driven landlords to discriminate against immigrants. It is not at all clear how an influx of immigrants can change the housing market. On one hand, the demand for rental units would go up but on the other hand, there may be displacement of native people in the local area. Such opposite forces are complex to analyse. Two main issues in identifying the net effects: ability to untangle the unobserved attributes that may be crucial in decision-making and ability to observe trends over a long period of time. Both are difficult to achieve. Regardless, demand shocks are generally dealt with by reciprocal and appropriate supply-side actions. Work by d'Albis et al. (2018) has identified supply-side responses. As noted in earlier chapters, supply is not easy to use as a policy device as it takes significant time and effort to scale up due to the nature of the market and processes that need following in completions. Therefore, inelasticity of supply in the short run greatly exacerbates the impact of the demand shocks. This inevitably leads to slow adjustment of the market. In markets where the built environment is highly regulated, the adjustment can be very slow. Market cycle can also play a significant role in the adjustment process. For example, the recessionary time period can be much slower than the expansionary time period. Such cyclical patterns should also be factored into the evaluation of policy measures. Several authors have commented on this stock adjustment issues. For example, Ohtake and Shintani (1996) wrote on the role of long-run housing elasticity and demographic changes on housing stock. More interesting, the development market may react to the adjustment issues differently and position the business proposition accordingly. That means stock may actually change over time, albeit gradually and perhaps quite slowly. The tenure mix and type of property and physical attributes may change. Kemeny (1995) noted that the unitary rental market (social housing competing with private renting in an unregulated environment) may have a different impact compared with more regulated dual rental systems. Such differences can influence the development market reactions. This is evidenced by Skifter Anderson et al. (2016), noting that a segmentation between rental and homeownership exists. This implies that immigration influx can lead to increases in housing demand, to be absorbed by rental stock adjustment. Although social housing can be positioned as a solution, it is important to note that access to social housing and the information barrier can play a significant role in whether the tenure becomes a viable option for the immigrant cohorts.

Academic studies have also indicated that the human capital endowment of immigrants such as education, work experience, ability to generate income and ability and willingness to invest in assets including housing can explain a significant part of the variation in housing outcomes among the natives and non-natives. Natives and non-natives' housing tenure choices are also determined by locational and housing effects such as the distance to the city/town centre, dominant housing tenure in the neighbourhood through neighbourhood effects (social, rental and homeownership), type of housing (flat, maisonette, detached), number of rooms and number of floors. Hall and Greenman (2013) found that satisfaction with a neighbourhood, housing quality, crime rate, safety, services and infrastructure deficiency are important determinants. Zorlu et al. (2014) indicated that neighbourhoods with high immigrant rates often tend to have low ownership rates, and slow increase in ownership. Studying American cities, Borjas (2002) also suggested that locational decisions made by native and non-natives may have determined the housing outcomes for both groups, which have resulted in ethnic enclaves and also could indeed increase the probability of homeownership for immigrants due to ethnic networking effects.

Ethnic networking can lead to spatial concentration due to similarities in bid rents. Especially when information is a key factor for successful settlement for newly-arrived immigrants, spatial concentration is important. Skifter Andersen et al. (2016) reported a

128 Demographic shifts and housing

strong correlation between the spatial patterns and housing tenure for immigrants. Their study of Helsinki and Oslo indicate that the two cities have the most uneven distribution of social housing across neighbourhoods with a high level of segregation of private renting and cooperatives in Copenhagen. This aspect may have been driven by the concentration of these tenures in the older housing stock in central Copenhagen. In Helsinki, owner-occupied single-family housing appears to be more unbalanced. The rented housing and owner-occupied flats are more evenly distributed across neighbourhoods. It is important to note that Helsinki has pursued a policy from the 1970s to enable mixing of different tenures, which can help with creating more balanced communities.

Several countries have been studied – for example, Gonzales and Ortega (2012) studied the Spanish experience; Stillman and Mare (2008) provides evidence from New Zealand, Akbari and Aydede (2012) examined the Canadian experience; Mayda (2006) used two sources of data – first, the International Social Survey Programme (ISSP) using the ISSP (1995) National Identity module (ISSP-NI) covering more than 20,000 respondents from 22 countries, including the United States, Canada, Japan and several Western European countries. The survey also includes a few Eastern European countries and one developing country, the Philippines. Mayda (2006) complemented the results from this source with the findings based on the third wave of the World Value Survey (WVS), carried out in 1995–1997. The WVS dataset includes more than 50,000 respondents from 44 mostly developing economies. A number of studies (e.g. Borjas, 2002; Zorlu et al., 2014) have noted that taste, preference, social network effects, household factors and factors specific to the country of origin in terms of attitude towards homeownership can explain a significant part of the variation in housing tenure outcomes. Aslund (2005) reported a lower probability of homeownership in a home with a male family head. Kuebler and Rugh (2013) documented a variation in homeownership differences among Whites, Asians, Mexicans and Cubans and attributed the differences to socio-economic status. Zorlu et al. (2014) found that Moroccan immigrants seemed to have a higher homeownership rate than the Turkish immigrants in the Netherlands. Interestingly, they suggested that the differences could be due to the immigrants' perception of homeownership shaped by the patterns in their home country along with other individual characteristics, household characteristics, family structure, marital status, parental background, neighbourhood, financial awareness and exposure. Age, gender, race and ethnicity have also been suggested as having significant influences on tenure choice (Goodman, 1990; Coulson, 1999; Nygaard, 2011; Zorlu et al., 2014; Skifter Andersen et al., 2016). In a recent study, Olayiwola, Nanda and Milcheva (forthcoming) analysed the housing outcomes of natives and multiple generations of non-natives using a longitudinal survey data in UK (Understanding Society). They find that probabilistic models for housing tenure choice show significant variation in the outcomes which are robust to several econometric specifications.

The link between migration and housing can be understood by analysing bid-rent functions carefully. As discussed in earlier chapters, bid-rent functions capture many consumer choice parameters that can shed important light on housing outcomes, especially locational aspects. The shape of the bid-rent function varies across households and economic entities depending on the income elasticity of commuting cost and demand for land. For the residential sector:

- Bid-rent reflects tastes and preferences – thus shapes communities.
- Bid-rent functions capture positive and negative feedbacks to the household choices.

- The factors that positively influences the residential bid-rent are: access to downtown employment, shopping and other urban amenities like schools, health services, fire, police services.
- The negative factors include crime, environmental pollution, congestion, noise pollution, taxes (given a level of public services).
- Different types of households exhibit different bid-rent functions:
  - E.g. Single adults or couples without children may make relatively high bids for land located near the downtown area but may make low bids for suburban areas.
  - However, families with children will bid relatively higher for spacious and safe suburban locations.

Therefore, the analysis of bid-rent functions can lead to a better understanding of the choices made by the migrant groups. However, the empirical estimation is heavily data-driven and much care is needed in the empirical analysis to be able to avoid pitfalls of endogeneity, reverse causality and unobserved heterogeneity in the econometric set-up.

Case study-I: Migration in the United Kingdom

The United Kingdom has experienced a significant amount of migration over the decades. Over the years, the UK public policies have increasingly started incorporating measures and guidance for achieving a more mixed and diverse communities. A strong assimilationist approach with emphasis on nurturing mixed and diverse communities has been adopted, which can help in achieving much-needed social cohesion and community wellbeing. In terms of housing and economic outcomes, some stylised facts from the report produced by the Migration Observatory are[3]:

- The foreign-born population has significantly lower ownership rates (42% were homeowners in the second quarter of 2017) than the UK-born (69%). In 2013, the foreign-born population had also significantly lower ownership rates (43% in 2013) than the UK-born (69%).
- The foreign-born population is almost three times as likely to be in the private rental sector (41% were in this sector in the second quarter of 2017), compared to the UK-born (15%). In 2013, the foreign-born population was also almost three times as likely to be in the private rental sector (38% in 2013), compared to the UK-born (14%). UK-born and foreign-born individuals living in London tend to have lower homeownership rates, and a higher likelihood of being in the rental and social housing sector relative to the rest of the UK.
- Recent migrants (i.e. those who have been in the UK for 5 years or less) are almost twice as likely to be renters (80% were in the private rental sector in the second quarter of 2017), compared to all migrants. Those migrants who have been in the UK longer tend to have accommodation similar to that of the UK born.
- UK-born and foreign-born individuals have similar levels of participation in social housing (about 16% of UK-born individuals and 17% of foreign-born individuals were in social housing during the second quarter of 2017). Compared to 2013, not much has

---

3 http://www.migrationobservatory.ox.ac.uk/resources/briefings/migrants-and-housing-in-the-uk-experiences-and-impacts/#kp1

changed in this regard. UK-born and foreign-born individuals had similar levels of participation in social housing (about 17% of UK-born individuals and 18% of foreign-born individuals were in social housing during 2013).
- Recent evidence suggests that immigration decreases house prices in England and Wales. This impact is likely to be the result of UK-born residents moving out of the areas with a large migrant concentration.
- New migrants tend to live in poor quality housing and neighbourhoods that are often home to other migrant communities (Heath and Chung, 2007). This can be due to ineffective and inefficient information network among the members of the migrant communities. A pertinent question can be asked: is neighbourhood networking good or bad for the migrants? Much of the answer may depend on the migration life cycle and time in the destination. However, lack of opportunity and significant mobility constraints can lead to patterns in deprivation and segregation, which has been well-documented in the literature.

Figure 7.2 shows the types of accommodation for UK-born and foreign-born individuals in the Labour Force Survey (second quarter of 2017). Foreign-born individuals seems to experience much lower homeownership rates (42%) compared with the UK-born's 69% homeownership rate. This also relates to the fact that foreign-born individuals are more likely (41%) to rent in the private rental sector, compared to the UK-born (15%).

- New migrants often enter the UK housing market in the least desirable housing – frequently in disadvantaged areas or where demand for housing is lowest. With increased migration resulting from the expansion of the European Union (EU), more new migrants have been seeking housing in or around rural areas, where employment in agriculture, tourism and related industries has been available to them. Such trends may reverse with the UK leaving EU in 2019 onwards. Immigrants often depend on housing tied to their employment and the private rented sector, where they may experience overcrowding, poor conditions and insecurity.

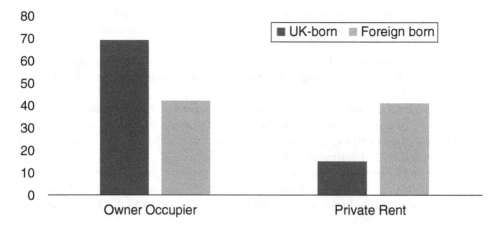

Figure 7.2 Accommodation of foreign-born in the UK.

Source: Based on data from Migration Observatory report, which is based on ONS Labour Force Survey data. http://www.migrationobservatory.ox.ac.uk/briefings/migrants-and-housing-uk-experiences-and-impacts

Demographic shifts and housing 131

- The increased migration to the UK can indeed put significant pressure on the provision of affordable housing, as there is already a significant shortage of affordable housing and at the same time, the demand for low-cost housing has risen very fast due to the widening gap between house price inflation and wage inflation across much of the UK regions.

From the mid-1990s to 2010, immigration has picked up significantly in the UK. The mix of tenure has remained predominantly skewed towards private renting for the recent migrants who are less than 5 years in the UK.

- One of the biggest risks to such sub-optimal economic and housing outcomes is deteriorating social cohesion and rising tensions and greater risks of conflicts among communities. Poor housing, competition for housing and high levels of population 'churn' have a detrimental impact on relationships at the local level between different groups, particularly in areas not used to such change.
- This is a very highly politicised subject with continuous interventions by political outfits and lots of rhetoric that may not always be based on facts. Highly politicised and negative debates about migration can result in whole groups of people, such as asylum seekers and migrants from EU accession states experiencing discrimination. These debates are often not based on evidence – there is a lack of basic information about new migrants, their numbers, their housing and economic circumstances, their status and their rights.

Case study-II: Migration in the United States

Figure 7.3 shows the distribution of housing tenure for natives and non-natives in the US in 2011. Homeownership for natives is almost 15 percentage points more than that of the foreign-born population. Naturalised citizens, though, have similar homeownership rates to

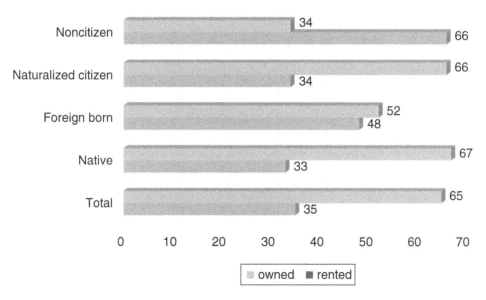

Figure 7.3 Distribution of housing tenure for US natives and non-natives (2011).
Source: Based on data from US Census Bureau, 2011; Trevelyan et al., 2013.

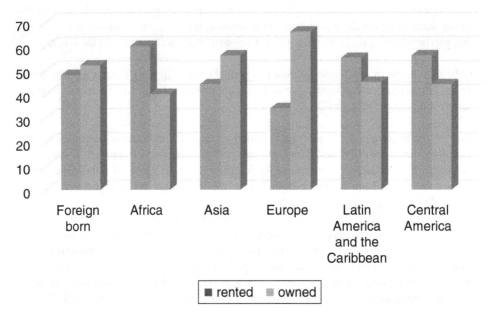

Figure 7.4 Distribution of housing tenure for US non-natives by region of birth (2011).

Source: Based on data from US Census Bureau, 2011; Trevelyan et al., 2013.

natives. It is important to note that it takes several years for the foreign born to attain citizenship in the US and as a result of that, their preparedness for homeownership may be better. Within the foreign-born population, people from Africa has the lowest homeownership rate, followed by people from Latin America. Migrants from Asia and Europe do have better homeownership rates (refer to Figure 7.4).

Figure 7.5 provides a more granular picture of the distribution of tenure by country of birth. The Dominican Republic shows the lowest rates of homeownership (almost 30% or lower), while people born in Germany, Canada and the United Kingdom show much higher homeownership rates (more than 70%). Citizens from India and China have moderate homeownership rates (55% and 60% respectively).

It is also seen that the tenure rates differ significantly by period of entry. It is not surprising that more recent years' (2000 or later) entries may show much less homeownership compared to the older generations due to their financial and economic preparedness in the US.

Policy concerns

The scale of migration, as defined by international and inter-regional levels, is very high across the world. Almost all countries usually devise policies to deal with various issues related to migration. Developed countries who usually receive a high level of immigration tend to devise policies that restrict easy movement of people from outside the country. Developing countries often devise policies to deal with inter-regional migration issues.

The key policy concerns related to the housing market are: First, spatial distribution of migrants – migrants' demand for housing leading to house price inflation and worsening

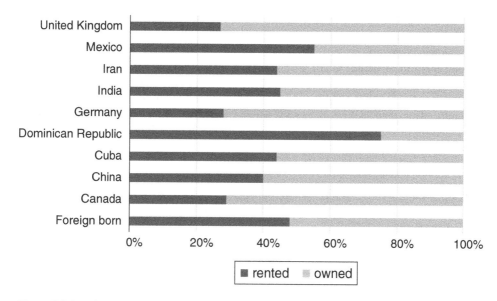

Figure 7.5 Distribution of housing tenure for US non-natives by country of birth (2011).
Source: Based on data from US Census Bureau, 2011; Trevelyan et al., 2013.

affordability for native/resident households. As discussed before, much debate has surrounded around the impact of migration on housing demand and thus housing affordability. Second, migrants entering social housing, reducing the supply available for domestic households. Third, social cohesion and inclusive prosperity in the face of pressures on public amenities which can compromise integration of the migrant population, leading to social tensions and lower level of societal prosperity.

The above policy concerns can be tested through various economic and rigorous econometric modelling. However, modelling the issues require identification and recognition of inter-related attributes that often lead to estimation biases. There are several factors that need to be considered in such modelling exercises.

Key factors to consider while modelling the dynamics of migration:

- **Labour market is the key:** Wage differentials are very important in international migration. Much of the international migration is driven by job market attractiveness and better salary with a close match to skills.
- **Information asymmetry:** Newly arrived immigrants are less likely to have the necessary information to make an informed judgement about house price differentials. Therefore, temporal variation, in terms of the start of the migration life cycle, can be a significant factor in understanding the migration impacts.
- **Migrant networking may hold it together:** It may reduce transaction costs and the information gap in housing and labour market outcomes. This can especially be of significance for less-skilled migration and migration that is driven by information network integration with origin and destination countries.
- **Severe credit constraint:** Many migrants are more likely to experience credit and mortgage constraints right away. First, they may have less constrained access to familial

capital. Second, due to displacement and lack of integration in global personal credit histories, they may not have adequate credit history to qualify for better credit products and services. It may take several years to be able to build enough credit history for favourable borrowing conditions.
- **Long-term motivation:** Often, migration is related to temporary stay/arrangement for a few years). As a result, housing market outcomes are of less importance and migrants may be more motivated towards higher savings rate for sending money to the origin country. This may lead to increased savings with higher MPS and potential higher demand for investment (possibly in housing market) in the origin country.

A model explaining spatial distribution of migrants[4]

Let's look at a modelling exercise. Different regions of a country receive different extent of migration. For example, in the UK, the Greater London area receives the most migration compared to other parts of the country. What factors can possibly explain such spatial variation? One easy explanation is labour market dynamics, i.e. availability of jobs and higher wages compared to other regions. Regional wage differential and social or ethnic network effects are important determinants, as they would capture expected utility stream differential in location j and the ability of individuals to draw on ethnic networks to minimise transaction costs in the housing and labour market. This can further help with lowering housing costs. Suppose the regional share (M) of migrants from country c in region j can be modelled as:

$$M_{jc} = \ln\left(emp_j/emp_c\right) + \ln\left(w_j/w_c\right) + \ln\left(pop_{jc}\right) \\ + \ln\left(hp_j/hp_c\right) + law_j + bilateral_{ic} \tag{7.1}$$

The regional share of immigrants from country h within region j can be defined as a function of the difference in the probability of finding employment (emp), relative wage differentials (w), differences in real house prices (hp) between region j and country c and strength of the network of citizens of country c (pop).

Equation (7.1) can be estimated with historical data with cross-sectional variation. For example, state-level variation in migration related to housing market in the United States can be explained by such modelling framework. In the model, there are several unobserved factors that need careful modelling. For example, immigration policies, laws governing immigration practices and regime can significantly explain the variation, which (*law*) can be built into the model. Note that such regulatory backdrop can also change over time, requiring modelling temporal aspects. Bilateral relationships (*bilateral*) between the origin and destination countries can also significantly affect the migration flows.

Another big question that has often been raised in the literature is: what is the impact of immigration on the demand for owner-occupied housing?

---

4 For further reading on this modelling framework see https://www.gov.uk/government/uploads/system/uploads/attachment_data/file/6354/1774894.pdf

A model explaining the impact of migrants on the demand for housing

- Several studies have looked at this issue (refer to the discussion earlier in this chapter). (necessary condition) – propensity of immigrants to form separate households relative to the departing and resident population, i.e. ethno-cultural difference might induce systematic variation in the likelihood of establishing a separate household.
- (sufficient condition) propensity of immigrant to become owner occupiers: preferences, immigrant self selection, discrimination and filtering policies may induce systematic differences in the likelihood of immigrants to own a house.
- Both can be modelled as a discrete choice function:

$$P\left[hr_{it}, ten\right] = \beta_0 + \beta_1 X_i + \beta_2 E_i + \beta_3 S_i + hr_{i,t-1} \\ + yr + \left[own/rent\right] + v \qquad (7.2)$$

where X, E, S, are socio-economic/demographic, economic and ethnic and country of birth vectors, respectively; and hr_{t-1} is headship status in the previous period, yr is length of residence in the destination country and *[own/rent]* is the user cost of housing.

Concerns of affordability

With high levels of movement of people through various types of migration, the demand for various types of housing goes up. As discussed before, demand for housing going up will lead to price increases when supply is severely constrained or cannot be scaled up to match the demand increases. Income levels across many countries have not grown at a high level for several years. Such a mix of dynamics will generally lead to worsening of affordability, i.e.

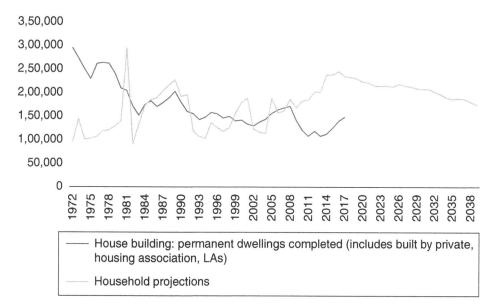

Figure 7.6 Housing building and household projection in England.

Source: Based on data from MHCLG, UK.

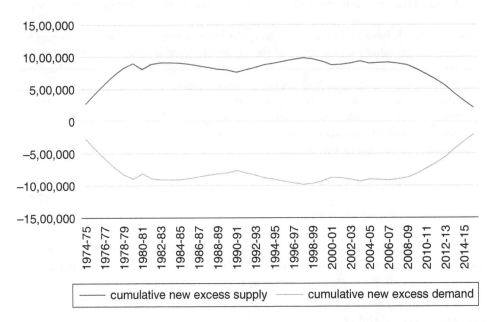

Figure 7.7 Demand-supply mismatch in England.

Source: Based on data from MHCLG, UK. https://www.gov.uk/government/statistical-data-sets/live-tables-on-household-projections; https://www.gov.uk/government/statistical-data-sets/live-tables-on-house-building

households with average income will find it difficult to be able to purchase an average priced house. The UK housing market provides a good example of this scenario.

Figures 7.6 and 7.7 show the demand-supply gap in England over a long period of time from 1974–75 to 2014–15. For the purpose of demonstration, I use the difference between the number of new houses built as new supply and number of new households as new demand. The cumulative new excess supply line is a mirror-image of the cumulative new excess demand. The chart shows a remarkable pattern in new demand-supply mismatch over almost four decades in England. Since 2005, a steady pattern of decline is noted in the amount of new excess supply.

Positive net migration sustained over a long time period has fuelled much of the overall household growth in England, as shown by Figure 7.8. While such growth has certainly helped the economy achieve strong performance, it has put strains on the housing affordability. The long term projection of population shows such trends to continue unless checked by migration and housing policy measures.

Coupled with dynamics from the existing housing units, it has put enormous upward pressure on house price levels as is evident from Figure 7.9. This will show different patterns across English sub-national markets.

Price inflation can be especially problematic for the households if it is paired with slow income growth. Figure 7.10 shows a very significantly widening gap between increases in median house price compared with increases in median gross income in England. A comparable analysis can be easily made on any other countries using these metrics. A sustained pattern such as this can create a massive decline in affordability for a large number of households. However, the national figures such as in Figure 7.10 can mask significant

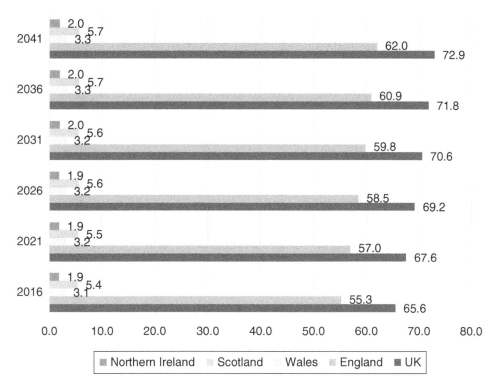

Figure 7.8 Estimated and projected population in millions of the UK and constituent countries, mid-2016 to mid-2041.

Source: Based on data from ONS, UK.

Figure 7.9 Demand-supply mismatch and price movement in England.

Source: Data from MHCLG household projections model. https://www.gov.uk/government/statistical-data-sets/live-tables-on-household-projections; https://www.gov.uk/government/statistical-data-sets/live-tables-on-house-building

138 *Demographic shifts and housing*

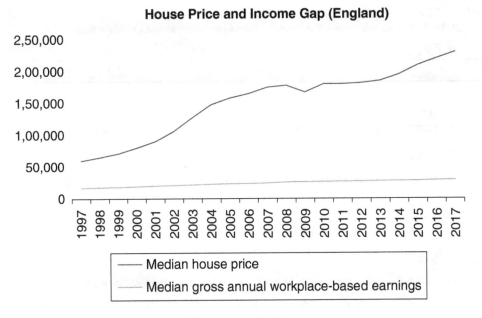

Figure 7.10 House price inflation and income growth (£).
Source: Based on data from MHCLG, UK.

spatial variation within the country. For example, when the metrics are compared across nine English regions, the regional variation is remarkable with northern regions showing better parity between these two metrics (house price and income), while the southern and eastern regions (e.g. London, South East) show a very severe level of gaps. An obvious factor is location of jobs and businesses being concentrated in those regions leading to much higher housing demand along with severe supply constraints. Such spatial variation can be easily found in other countries. Rebalancing of regional economies through policy measures is required over a long term to address such patterns.

As discussed in an earlier chapter, a frequently used metric for evaluating affordability is the affordability ratio, which is computed as the ratio of house price and income (usually medians). Lenders use it for granting loan amount. A score of 4.0–4.5 is usually seen as affordable. Referring to an earlier discussion, this ratio has changed significantly across many regions of England due to much higher house price inflation compared to wage inflation. Figure 7.11 shows how the metric for three key regions of England has changed over time; it almost tripled over two decades.

Such changes are also evident at a more local level. Figure 7.12 shows four major local areas in the Thames Valley region. In just two decades, the affordability ratio changes dramatically – in Reading (2.3 times); Slough (3.3 times); West Berkshire (2.2 times); Wokingham (2.5 times).

With lenders not willing to lend more than 4.5 times of the income level, it becomes unaffordable for a large number of households. There have been reports of more reliance

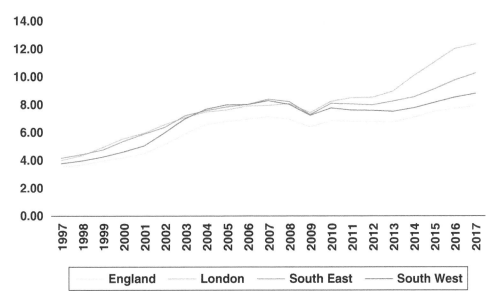

Figure 7.11 Affordability ratio across English regions.
Source: Based on data from MHCLG, UK.

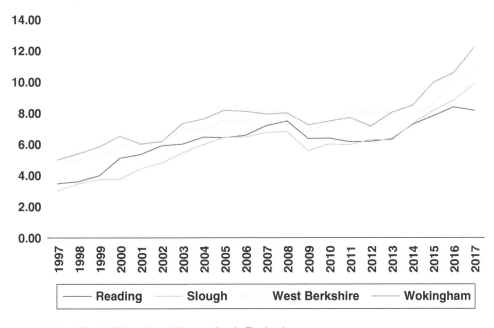

Figure 7.12 Affordability ratio within a region in England.
Source: Based on data from MHCLG, UK.

140 *Demographic shifts and housing*

on wealth, past savings, help from parents and relatives[5], help from government schemes (e.g. help-to-buy scheme in England introduced in 2013) and other source of funding.

Several policies have been rolled out to help buyers and tenants over the past several decades in England. Examples of such policies are – Right to buy, Stock transfer, s106 affordable housing provision, Buy-to-let, Shared ownership, Help-to-Buy, Changes in Stamp duty Land Tax (SDLT), to name the significant ones. However, due to severe supply constraints (with supply curve being stiff in the short-run as well as in the long run), the housing polices have not quite achieved the desired levels of success.

There is also another fundamental problem. Referring back to the demand equation (3.13) in chapter 3, equilibrium prices can be determined by the price equation that incorporates the demand and supply-shifters as follows:

$$(p_H/p_C)^* = f\left(I, r, Cr, W, tH, pH^e, N, h, m, H_E^s, \pi, cc\right) \tag{7.3}$$

If we were to estimate equation (7.3) with real world data, we may express it in a log-linear format and estimate it as a regression equation. A proportionate rise in both the housing stock and the number of households can leave the real house prices unchanged. However, housing demand is constant if the income elasticity = 1, for given values of the other variables. However, if income elasticity is greater than 1, that also implies that the income elasticity of housing demand is higher than the price elasticity. Therefore, the affordability may deteriorate over time unless the supply of housing rises much faster than the number of households and other market stabilising effects, like changes in interest rates or other credit market constraints. The English housing market provides abundant evidence of such affordability dynamics.

In view of the severity of the affordability problem, the UK Government had set up a major review of the constraints in the housing market. The Barker Review of Land Use Planning (2006) provided a number of recommendations[6]:

- The planning system is the main cause of housing shortages and weak supply elasticities.
- The planning system should be made more sensitive to market conditions (extra land should be released when prices are rising).
- Affordability targets should be adopted at national and regional levels.
- A Planning Gain Supplement (PGS) should be introduced as a percentage of the increase in land value resulting from the grant of permission so that the society gains from an uplift in property values.
- Establishment of a new National Housing and Planning Advice Unit (NHPAU) on methodology.

In short, the main conclusion from the review is that the planning process adds significant constraints. The UK planning system is well-developed, evolved over many decades and is rather complex. In a very brief note, the central government makes detailed population projections by various demographic parameters such as age, gender, marital status, etc. For

---

5 Often referred to as Bank of Mum and Dad in popular media, based on a report by Legal and General. https://www.legalandgeneralgroup.com/media/1077/bomad_report_2017_aug.pdf
6 https://assets.publishing.service.gov.uk/government/uploads/system/uploads/attachment_data/file/228605/0118404857.pdf

each group, trends in 'Household Representative Rates' or Headship rates are calculated. Multiplying together gives household projections. These computations feed into Regional Spatial Strategies, which have influence on local planning frameworks and thus the number of planning permissions.

However, there are several issues with the British planning system and how the targets are made and the strategies are informed. The factors affecting household formation and households' housing preferences (which are important factors determining the demand for housing) are not fully considered. Even if the number of new households are matched with dwellings and forecasts are largely accurate, the rising level of demand from existing households would still produce a large shortage, due to income elasticity being greater than price elasticity. Lengthy planning processes can raise time to pass through the development control, uncertainty regarding approval of projects and other development costs (such as impact fees). Such inefficiencies may affect the housing production process and also the intensity of innovation and the structure of the industry. Moreover, planning delays, uncertainty and other inefficiencies make transaction costs higher and create barriers to entry for the new entrants. The conventional metrics like the growth rate of house prices, the price-to-rent ratio, and the price-to-income ratio can be misleading because they fail to account for the time series pattern of real long-term interest rates and the predictable differences in the long-run growth rates of house prices across local markets which can be captured more effectively by the User Cost of Housing model discussed in an earlier chapter[7].

The concerns of affordability are particularly of importance to young households as they face more severe income constraints along with weak savings and wealth profile. In the next section, I turn to a discussion of the housing outcomes for young people. While the direct reference is made to that of England, the lessons and issues can be similar in other countries.

Housing for young people

Housing choices and ability to get onto the housing ladder have changed over the years across all age groups. Two most notable aspects, as discussed before, are: house prices (both nominal and real terms) have risen very sharply in the UK; and income levels have either risen only modestly or even remained flat. This has resulted in widening house prices to the income gap. One of the biggest fallouts of this has been much larger deposit requirements. For example, the house that used to cost £150,000 in the mid-1990s is more than doubled to £300,000 in 2018. A commonly applied deposit requirement of 15% implies that a household would require £45,000 as a deposit. Such a scenario has hit the younger generations hardest, as they have much tighter financial ability to be able to pay such a high deposit. Moreover, as we have seen before, financial institutions usually apply 4.0–4.5 as multiple of income as a maximum loan amount. At the same time, from Table 6.3 in Chapter 6, we have seen in most regions in England and Wales that the income multiple far exceeds 4.5. For a large number of fresh university graduates with only a modest level of salary (£25,000–£35,000), they can only afford a limited amount. Both high deposit amount (due to high house price) and inability to qualify for high enough loan amount have become a double-edged sword for a large number of young British households, for whom the private renting sector (PRS) is perhaps the only viable option although the rental growth has also made that option increasingly

---

7 "Assessing High House Prices: Bubbles, Fundamentals and Misperceptions" 2005 (Himmelberg, C., Mayer, C., Sinai, T.) *Journal of Economic Perspectives* 19:4 67–92.

unaffordable. Several estimates show a very significantly high number of young households will have to live in PRS housing.

Young families, especially with low income level, are getting marginalised in the UK. The trend is leading to much instability in younger age groups and is not sustainable. Relying on PRS is not a very good option with the current status of PRS. As we have seen in Chapter 6, PRS in the UK has several weaknesses, especially insecurity of tenure. Without a much more aggressive policy framework, it will not be possible to change the state of the affair towards making PRS as a very viable option.

The main objective of the British Household Panel Survey (BHPS) data is to further understanding of social and economic changes at the individual and household level in the UK. Starting in 1991, a total of 18 years of panel data have been collected by 2009. In a detailed report, Clapham et al. (2012) described and analysed several areas of issues and challenges for young people in the UK in terms of housing. Using British Household Panel Survey (BHPS) data, they identified nine pathways, as per their description below, based on several characteristics – in terms of tenure, household type, income and education.

Pathway 1: stay at home to own: this is the most common pathway among young people aged 18–30 years. Due to rising housing and education costs, many young people are choosing to remain in their parental home for longer until they have built up sufficient financial savings to make a first-time purchase and/or are able to form a family on their own.

Pathway 2: dual income, no kids owners (DINKOs): this has also been found as a common housing pathway. These young people enter the Ppivate rented sector (PRS) housing with their first jobs, after leaving the parental home. This pathway may also involve time spent on living in the PRS either as a single-person household or with unrelated adults such as friends, colleagues, flatmates, etc. This is somewhat similar to Pathway 5 below – the young professional enters the pathway with the difference that this pathway has a higher percentage of forming families and subsequently opt for enter into owner occupation.

Pathway 3: two parent families: After their stay in the parental home, these young people enter either owner occupation or the PRS, with the majority being owner-occupiers after 10 years. Unlike Pathway 4 (the early nesters pathway) below, these young people in this pathway stay longer in the parental home. This may also indicate that they take longer in terms of forming families.

Pathway 4: early nesters: These young people form their own families, their own 'nests', after leaving the parental home by the age of 21. Many people in this pathway enter owner-occupied homes with others in the PRS. The authors note that exits from the parental home are often a result of unplanned pregnancy, sometimes in the presence of significant family support to facilitate their move.

Pathway 5: young professional renters: In this pathway, young people leave their parent's owner-occupied home often for higher education and then onto PRS for pursuing early professional careers as new graduates.

Pathway 6: in the social queue: This is for the social housing. Their parental home is in the social rented sector. A large number of young people may stay in the parental social rented home. Those leaving the parental home may get into their own tenancies in the social rented sector or the PRS. Compared with Pathway 1, these young people tend to have slightly more couple or family-formation after 10 years.

Pathway 7: lone parents: This is a very clearly dominated pathway by women, making it the only 'clearly gendered' pathway. These young women leave the family or parental home

Demographic shifts and housing 143

and enter social rented housing, often soon after or just before having a child. Interviews by the authors indicated that much dynamics is determined by patterns of relationship formation and breakdown that are often short-lived. Some young people also enter into the PRS.

Pathway 8: social renting families: Similar to Pathway 6, these young people also leave their parental home and most of them get into their own social rented tenancy. A small number may enter the PRS. Many of these young people in this pathway are in wedlock with children. There is little tenure movement for these young people and they often find it difficult to have tenure independence and mobility. Clapham et al. (2012) note that

> accessing the social rented sector is difficult for young people in some regions of the UK, and consequently some young people have developed alternative access strategies. For example, some young interviewees reported declaring themselves as homeless, while others temporarily split with their partner in order to access social housing as a single parent.

Pathway 9: chaotic: The nature of this pathway is unstable, often punctuated by frequent changes in tenure and housing status as well as periods of homelessness. As a result, it is difficult to get a good information base from the British Household Panel Survey.

For a detailed discussion of these pathways and estimated numbers belonging in those pathways in 2008 and 2020 projections, refer to the Clapham et al. (2012) report.

The factors affecting young adult's decision to move or stay at home longer have been studied extensively in the literature. Family structure and situation, labour market pull, educational attainment, financial independence and social trends have been examined in detail and found to be significant drivers behind those decisions. In term of family structure, marital relationships, related issues, health conditions of parents and closeness to parents have been

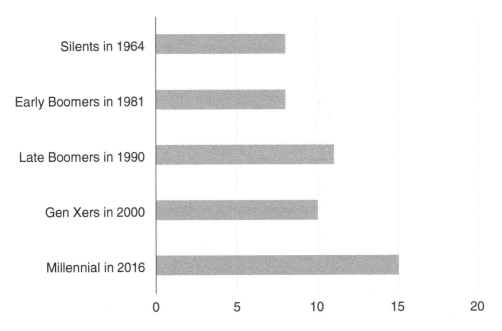

Figure 7.13 Generations likely to stay at parental home (% of 25–35 year olds living in parental home).
Source: Based on data from Pew Research Center.

144 *Demographic shifts and housing*

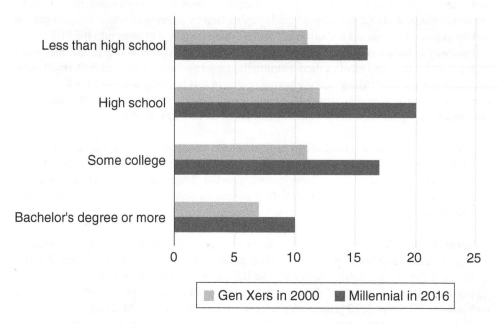

Figure 7.14 Staying at parental home by education attainment (% of 25–35 year olds living in parental home).

Source: Based on data from Pew Research Center.

found to be significant factors – young adult's likelihood of staying put in their parental homes increases if the parents are married or remains in wedlock. Single parent, presence of step parents, health problem of parents and strained relationship with the parent have been found to be the push factors for young adults to leave the parental home (Aquilino, 1990; Cobb-Clark, 2008; Burn and Szoeke, 2016).

Figures 7.13 and 7.14 show a very interesting pattern in preferences of the new generation.

Uncertainties with respect to labour market, economy and financial markets also play a big role in co-residence decisions. In the face of economic uncertainty and financial insecurity, being able to co-reside with parents is a good option for the young adults as it allows them to ride through the bad times without much reduction in consumption level and life style choices. Preserving the consumption is an attractive aspect. Stone et al. (2011) found such links to be of significance and the importance has been highlighted more prominently in terms of the links with broader economic melt-downs such as the Global Financial Crisis of 2008.

Education is a very important aspect of young adults' decision to co-reside or not. Stone et al. (2014), Burn and Szoeke (2016) and Matsudaira (2016) show the significance of education and labour market outcomes on this issue. Several aspects are note-worthy: first, higher education can delay family formation and partnership; second, much larger number of women opt for higher education these days compared to a few decades ago, which has resulted in higher number of female graduates coming back to home; third, cost of higher education has become very high with significant loan amount to be repaid as part of the student loan. A low initial salary level and the housing cost (of renting) in job locations may prompt young adults to co-reside in their early years of their professional career. At the same time, lack of higher education can affect job prospects, which in turn may influence a decision to stay at a parental home. Referring to young adults in Europe, Burn and Szoeke (2016) argue that the low cost of living in Nordic countries can lower the likelihood of co-residence compared to southern European locations where the housing costs tend to be much higher.

Commenting on the co-residence aspects in Australia, Cobb-Clark (2008) provided a detailed account of various issues of co-residence. One of the significant aspects from economic perspectives is the possibility of using co-residence as a mechanism for resource-transfer between generations. This poses several interesting problems in terms of the timing of such transfer, post-transfer relationships and life satisfaction for both the transferor and transferee as well as a plausible link of cultural background and norms governing such decisions. As the author pointed out, in most western developed countries, the flow of resources originate mostly from parents to their adult children and the decisions about leaving home are jointly undertaken by the parent and the leaver. Culture and social norms, pressure or triggers from the social network can influence such decisions in a significant manner. Cobb-Clark (2008) also documents literature on life satisfaction due to such movements. For example, parents whose nests are empty appear to have higher marital satisfaction (Clemens and Axelson, 1985; DaVanzo and Goldscheider, 1990; White and Edwards, 1990). Countries do differ in this aspect – Italian and Spanish parents seem happy about such an arrangement while American parents are likely to be less happy when their adult children live with them (Manacorda and Moretti, 2005). Elasticity of housing demand and prices can also play a big role (Ermisch, 1999).

Clapham et al. (2014) have also highlighted a self-selection process in which dual income earners with no children, who have the highest education and income levels of all the young adult pathways, may actually choose to live in better quality private rental accommodations in desirable locations compared to opting for sub-optimal choices with the possibility of saving for subsequent homeownership. This marks another important aspect, i.e. recent social trends among the young adults. Since the advent of internet and digital technology, there has been a change in the young adults' perspectives on life cycle choices and consumption patterns. Typical or standard life cycle choices surrounding important life events such as marriage, timing for parenthood as well as a regular consumption pattern and more importantly short-run and long-run savings behaviours have evolved. A Merril Edge review (2017) (https://www.merrilledge.com/report) notes that there are now significant generational gaps in terms of attitude towards consumption and long-term savings. Recent generations are more likely to emphasise more heavily current life satisfaction than retirement planning. Penman and McNeill (2008) also report on higher likelihood and occurrence of non-essential, luxury and impulsive purchases. This change in social trends significantly affect the long-run financial planning such as retirement with much ramifications for financial needs of later lives when, due to medical innovations, recent generations can expect to live much longer than earlier generations. This leads us to the next section which is devoted to housing issues in the face of the ageing population.

Housing for old people

Socio-economic outcomes at old age will have cross-sectional heterogeneity that can possibly be explained, to some extent, by key economic and financial decisions made in younger life which, in turn, could possibly be attributed to variation in cognitive abilities, behaviour,

146 *Demographic shifts and housing*

perception and interactions with the external environment. Policy prescriptions will need to consider this heterogeneity and also need to reflect on what wellbeing means to different people.

Ageing populations across many countries will have very significant implications for the future of modern societies with much changes in resource demand and utilisation. The numbers are staggering in terms of scale, size and growth of ageing populations. The change is very fast. Just over the last few decades, there have been major changes, as the following figures suggest:

- World population has been projected to increase 3.7 times from 1950 to 2050, but the number of those aged 60 and over will increase by a factor of nearly 10.
- Among the elderly, the 'oldest old' – i.e. those aged 80 and over – is projected to increase by a factor of 26 (Bloom et al., 2011).
- About 37% of the European population is projected to be 60 or over in 2050, up from 20% in 2000 (DESA, United Nations).
- Within Europe, the countries with a large ageing population include Croatia, Finland, Germany, Italy, Spain, Switzerland and the UK (refer to Figures 7.15, 7.16).

This massive change in the structure of the population is of huge significance now and for many decades into the future. There is a natural limit to life expectancy. However, the current trend is upward – current younger cohorts can expect to live much longer than the previous generations. This means population is not only rising but also living longer. This can potentially change the economic and social dynamics across the world, possibly leading to reorientation of economic power. Policies need to be made, implemented fast to tackle the challenges. While there are substantial negative implications, there are also significant positives and opportunities. As one of the reports from the Institute of Fiscal Studies (Banks et al., 2016) note,

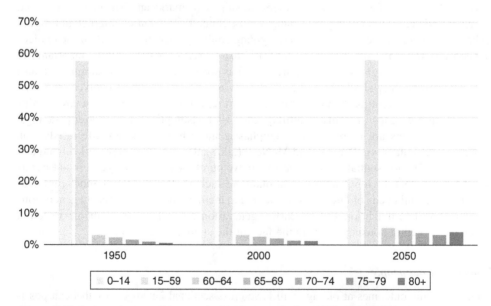

Figure 7.15 Scale and geographic spread of ageing population, 1950, 2000, 2050.

Source: Based on data from Population Division, DESA, United Nations: *World Population Ageing 1950–2050*.

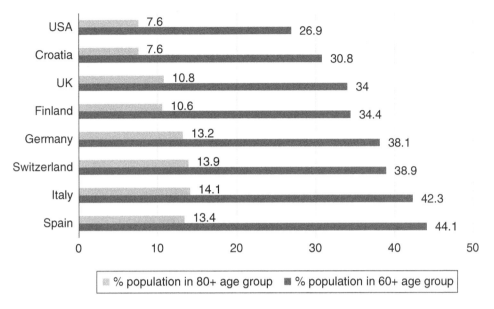

Figure 7.16 Scale and geographic spread of ageing population by country.
Source: Based on data from Population Division, DESA, United Nations: *World Population Ageing 1950–2050*.

narratives of these upward trends in survival are often couched in terms of societal burden when they should in fact be celebrated: the increased longevity of the population is one of the greatest achievements of the modern age, bearing testament to continuing improvements in public health and social care. The expansion of the age spectrum provides the potential to enhance opportunities for older people to make an even greater contribution to our society – one that has been hitherto poorly recognised – in supporting younger generations financially, practically and in the transmission of wisdom, in volunteering, and in active engagement with local and national political issues.[8]

The same report also note that "inevitably, these demographic changes also raise the need for complex, high level planning considerations in economic, health and social policy." They identify several areas of concerns and needs for strategic planning, for example, income security for older people, pensions and retirement benefits, social protection, the prevention of impoverishment and social isolation in old age, access to quality health care, effective and affordable social care, the promotion of age-friendly environments that enable independent living, the prevention of discrimination and securing the human rights of the ageing population. There are strong linkages of these with health implications due to age-related chronic diseases (e.g. cardiovascular disease, cancer, diabetes, dementias) and related human life condition states (e.g. cognitive, sensory and physical impairment).

8 Banks, J., Batty, G.D., Nazroo, J. and Steptoe, A (2016) The dynamics of ageing: Evidence from the English Longitudinal Study of Ageing 2002–15 (Wave 7). The Institute for Fiscal Studies. https://www.ifs.org.uk/uploads/elsa/docs_w7/ELSA%20Wave%207%20report.pdf

148 *Demographic shifts and housing*

There are several key issues having profound impacts on individuals, society and economy. The issues are inter-connected and often cross various spheres of human existence.

- *Economic issues* include financial security (e.g. appropriate pension plan), shelter security (e.g. suitable housing) and health security (e.g. appropriate health care and ability to pay for unforeseen and increased health expenditure).
- *Social issues* include the role of senior citizens in the society, ways of ensuring social cohesion in the face of unbalanced age patterns, etc.
- *Personal issues* include overall wellbeing, happiness, familial relationships, social networking, active ageing, etc.

The ageing population can cause several significant costs which can lead to impaired wellbeing.

- *human costs* as life in old age for an increasing number of individuals becomes challenging leading to impaired *personal wellbeing*;
- *social costs* as a large section of the population may end up remaining less engaged in social interfaces affecting *societal wellbeing*;
- *economic costs* due to rising health care expenses, inefficient allocation of public funds and less productive elderly population impeding country's *economic wellbeing*.

A particular need of human life is shelter, which tends to become specialised as we grow older. Increasing dominance of activity or mobility constraints in daily life, along with the aspirations to continue several aspects of younger life, lead to a very different need and demand for housing than the mainstream housing sector. As a result, the demand for bespoke housing and greater financial security catering are swelling and is expected to reach a staggering level in the foreseeable future, even if we believe in the most conservative estimate. A longer lifespan may call for behavioural responses or adjustments, for example, longer working life, change in the saving rate or contribution into retirement pension schemes. Such adjustments in economic behaviour can have a profound effect on the financial health, workings and aggregate productivity of an economy. It will also have significant impact on the structure of workforce, e.g. shift towards care services along with certain types of non-manual jobs, e.g. non-manual jobs becoming available to the elderly cohort. Such patterns in workforce structure can entail social implications and influence social cohesion.

There is now a substantial and growing body of research across many disciplines and cross-disciplinary studies on these issues. The implications for housing demand and supply are very important areas of concern. The demand for housing for elderly members of the society can fall into several categories depending on age profile and activity constraints. Health care implications and housing needs are intertwined for the group. Depending on the severity of the activity constraints, some may receive care in hospitals, care homes, nursing homes, retirement villages and residences or their own homes. Due to shortages and constraints on the hospital bed availability, more and more cases are appearing in favour of finding solutions for at-home care provision. If the resident is hospitalised, the person is then institutionalised. Within the cohort who remain out of healthcare facilities, they either remain in their existing homes or move to other age-friendly houses that are suitable to their taste and preferences for retirement life. The market for retirement housing offers these housing options. As the relative share of the ageing population is going up across many countries, there are various retirement housing options being constructed by developers and supported by policymakers.

An inadequate supply for the retirement segment currently persists across major markets. This is partly due to a lack of awareness of the size and nature of the potential demand. The tastes and preferences of the consumers of the retirement housing differ significantly with the mainstream housing consumers, which is highlighted later in the discussion.

While adaptation of the current residence can be a viable option, the facilities which are bespoke and purpose-built can offer greater quality of life and a more friction-free environment for elderly living needs. Environments that are conducive for elderly living require not only appropriate housing features but also adequate health and other care services for regular, routine as well as non-routine care needs for the elderly residents. Having specialised housing options for retirement living can be much better as it may limit the number of elderlies getting into more expensive hospital or other institutional arrangements. Demand for health and other care services can be higher if residents live in their homes which are adapted to suit their needs (Ball et al., 2011).[9] The issues and implications for the demand for specialised retirement facilities are manifold and requires analysis within appropriate contexts and framework (Litwak and Longino, 1987). The aspects of demand for retirement housing are a lot different from those of the standard housing market. The price and income elasticities are different due to different taste and preferences as well as the nature of the financial constraints that elderly typically face. As Gilleard et al. (2007) note, older people can be reluctant to move from their existing homes. Comfortability of own home environment and familiarity with existing and personalised housing services along with memories, social networks and neighbours and neighbourhoods are all significant deterrents for elderly residents to move out of their existing houses. Usually life changing events, movement of other near family members and unsuitability of the current home can act as 'triggers' for the decision to move. Ball et al. (2011) note that isolation, bereavement and activity-restricting health events are likely reasons for elderly citizens looking for alternative suitable housing.

Some studies have concluded that asset prices may fall as elderly cohorts sell off their assets (phenomenon of 'asset meltdown' and presence of 'income-poor and house-rich' households), e.g. asset meltdowns in housing markets due to decreased demand from the ageing members of the post-World War II baby boomers (Mankiw and Weil, 1989). The decision to liquidate home equity may take the intricate behavioural process such as 'human inertia' or 'procrastination' in stair-casing down the home equity ladder that cannot be reflected by the hard economic data. A major element of individual wellbeing is financial security. Financial security in modern life is often linked to housing asset accumulation as housing can be seen as a store of wealth and it is the most important channel of creating wealth. Especially at the old age, housing can be seen as providing an 'old age insurance' when the income stream becomes heavily limited and mostly stagnant. Moreover, a longer retirement life due to an extended life horizon may lead to greater reliance on home equity and higher propensity to consume out of the accumulated home equity.

Several papers (see review in Cuaresma et al., 2014) find positive effects of population ageing on economic growth. However, Acemoglu and Johnson (2007) and Lindh and Malmberg (2009) find negative or no effects of population ageing on economic growth. The analysis can be performed within a simplified two-sector economic framework with a retirement sector (mostly comprising the health care industry) and the non-retirement sector. The main

---

9 "Housing Markets and Independence in Old Age: Expanding the Opportunities" 2011 (Ball, M., Blanchette, R., Nanda, A., Wyatt, P., Yeh, J.). http://www.mccarthyandstone.co.uk/documents/research%20and%20policy/oorh%20full%20report%20may%202011.pdf

argument for a negative correlation is based on a possible shift of human capital towards the retirement sector due to the ageing population, which may adversely affect productivity and consequently lead to subdued economic growth. Aisa and Pueyo (2013) offer a theoretical model arguing that an increased rate of capital accumulation by the elderly cohort may offset (or even reverse) the negative effects due to weak substitutability of the factors of production in the non-retirement sector.

It can be argued that the positive feedback effects of capital accumulation may not be fully realised since such capital accumulation is typically achieved by the elderly cohort through building up the home equity over their mortgage tenure, and the accumulated wealth is largely locked into the house. The accumulated housing capital can only be released through an increased marginal propensity to consume out of the housing equity (i.e. housing wealth effect) and an effective 'recycling' of the elderly-occupied housing units through provision of appropriate retirement housing. These conditions are rather difficult to meet due to an inherent inertia to move into retirement housing, the prevalence of health safety nets and an acute shortage of retirement housing across all countries experiencing the ageing population and high dependency ratio.

Ball et al. (2011: pp. 16–17) have outlined several types of specialised housing for the elderly. The most intensive care is offered on an institutional basis in the form of residential nursing homes and specialised care centres. People with severe conditions or disabilities may require the highly specialised facilities. For the purpose of the housing market analysis, the segment that offers the environment suitable for independent living is interesting to study. These type of facilities offer a retirement experience with the provision of communal areas for social interactions, activity-based programmes for resident community engagement and other facilities to keep residents active and engaged with others. Ball et al. (2011) point out that medical studies show that independent living is an important requirement for health and general well-being. They also emphasise the roles of social capital, personal support and financing (Tang and Lee, 2011; Cannuscio et al., 2003; Pastalan, 1995). Mental health and happiness are important considerations.

In terms of house types, flats in apartment blocks are common and also terraced, semi-detached or detached properties in retirement villages. The most specialised retirement accommodation is the 'sheltered accommodation' for single people or couples, which are purpose-built blocks of flats having communal facilities or lounges, and full-time on-site staff and supervision, building management and support and remote assistance on demand in all other times (Ball et al., 2011). Another type of retirement living is 'retirement villages', which typically cater to residents with higher income and wealth. As Ball et al. (2011) point out, only 1% of English specialised retirement housing is in retirement villages, whereas in many other countries, retirement villages tend to have much greater shares, including in Australia and the US (Piggott, 2007).

The revenue generation model from the provider's perspective is different for specialised retirement accommodation. In England, specialised accommodation tends to be rented. Some providers also offer accommodation and care as a package with an appropriate fee structure. This tends to be the case for residents with care requirements. Healthcare considerations are very important and high levels of satisfaction are reported for owners (Ball et al., 2011). Kim et al., 2003 also note that while the lifestyle aspects are important motivations to opt for the option of retirement communities, the quality of health care facilities and services are also very important.

Although the above options are there, having the suitable arrangement at their own home seem to be the most preferred option. The inertia to move is very significant in the housing market. Being accustomed to their own home, neighbourhood and social network are important determinants of 'inertia' in the housing market. This is especially strong in the retirement housing market. Location of family members and friends nearby also adds to the inertia (Gilleard et al. 2007). Piggott (2007) also noted such inertia in Australia and elsewhere. However, such inertia tends to be broken with health and life events. Activity constraints and ailments that require specialised care and continuous supervision can be very detrimental for the aspirations of independent living. As people age, the incidence of such events become more frequent and severe. This is termed as the 'ageing with effort' effect which will increasingly affect housing choices (Ball et al., 2011). The medical innovation coupled with technology in the housing environment will attempt to alleviate as much as possible. As people age, mobility constraints start to influence housing and location choices at an increasing level. The ability to perform daily chores is a significant aspect of independent living. Climbing stairs, lifting objects at home and performing daily tasks become increasingly restricted and potentially dangerous as people age. At the same time, at that age, people tend to live on their own or with their partners who are also old. As a result, people feel insure at the face of the increasing likelihood and devastating consequences of physical injuries that can be sustained from falls and other domestic accidents. An increasing level of mobility constraints and deterioration of physical abilities can deter someone's favourite activities, e.g. gardening, morning walks, visits to local shops, etc. Moreover, maintaining and cleaning their own living space can also become problematic, which can add to the health woes and worries.

Deteriorating health conditions can be a very important decision-making factor for moving out of existing homes. Croucher (2008) points out that almost all moves were related to experiencing health problems linked with diminished mobility. While moving out is one option, adapting existing homes is also frequently undertaken. Ball et al. (2011) notes that the UK data show significant adaptations. Many households with a core member aged 75+ have adaptations related to hand rails (30% of them), bathroom modifications (26%) or alerting devices installed (14%), according to ONS data from 2010. However, there are limits to such adaptations and the more complex ones are expensive, so that moves to bespoke accommodation may become more sensible and cost effective. Moreover, the adaptation work may be disruptive over a long period, which can be problematic for the elderly members. Some studies (Litwak and Longino, 1987; Bradley, 2011) identify 'trigger' or 'inflection' points in the life cycle such as at retirement with physical constraints being minimal; with the moderate disability when specialised independent living is more desirable option; and when care homes is the only option with chronic and severe disability. Gilbert et al. (1994) points out an important factor when deciding to move family members and caregivers. They often take on the 'catalyst' role. The whole decision-making process may involve significant involvement and consultation with the family members and caregivers. The quality of health care services and proximity to family are influential factors. This can also be potentially contentious, as the elderly members' aspirations and wishes may run in contrast with those of their family members. Regardless, the health and care consideration may become the deciding factor.

Literature has identified lack of preparation for a vast majority of old people. Being myopic and not having the foresight of what lies at the elderly stage make people vulnerable to sudden changes. Surprising and unfortunate health and life events can come to anybody's life cycle and create the need and demand for retirement housing facilities. A number of studies (for example, see Sabia, 2008) points towards commonality in fundamental drivers across countries. The trigger events can be sudden and common (see Silverstein and Zablotsky, 1996). However, the suddenness and lack of preparation can be quite problematic as the potential solutions may not be well thought out and most likely be less ideal or sub-optimal. Therefore, planning of some sort is always welcome.

Isolation is very common across all elderly cohorts. With nuclear families, ease of mobility and globalisation, living alone has become common. For elderly cohorts, once they have raised their children to adulthood and taken them to the college-level education stage, the family homes become empty with over-consumption. Losing a spouse can be a very common reason for living alone at a later stage of life. Also, divorce rate and choice of staying single can be a reason for isolation in late life. Isolation can lead to issues for mental health and overall wellbeing. Moreover, for the physically-constrained elderly cohorts, isolation is not only a mental health issue but also can pose as a significant risk to their frail physical health. Ball et al. (2011) point out to the relevant statistics in the context of the United Kingdom, the number of older people living alone in the UK is increasing, with almost 50% of those aged over 75 currently living alone. Also, women are more affected by loss of spouse as their husbands, being men, have lower life expectancy and also due to the fact that husbands tend to be older than their wives. Widowhood affects a large number of older women, with over 60% of women aged 75+ being widows. The number of single elderly people is also expected to grow substantially in the future, with over six million of those aged 65+ expected to be living alone by 2033 in the United Kingdom only.

Numerous adverse health conditions and consequences can be triggered by loneliness and social isolation. In a number of studies on the ageing population, the link between loneliness/social isolation and ill-health have been reported. Survey and interview studies of elderly people frequently confirm such links as indicated by the respondents. There can be many ways of dealing with such issues, breaking the links as interventions that can lower the severity and extent of loneliness and programmes that can bring the socially-isolated people together can elevate the mental, physical and overall wellbeing. However, it is important to note that impacts of loneliness and isolation can have significant individual-level heterogeneity. Personal experiences can differ along with triggers. Different individuals react differently to the same interventions and triggers. Therefore, it is important that all intervention programmes work closely with the individuals. There is an interesting link between loneliness and activity constraints. The link is quite complex. At the simplest level, activity or mobility constraints are likely to contribute significantly to sedentary life or daily life being mostly confined in homes, which implies lot less interactions with the outside world and other people. Robinson and Moen (2000) note that the negative experience of living alone may be related to restrictions on personal mobility. Studies that report such correlations do need to be careful of unobserved heterogeneity that may exist at the individual level.

Understanding these issues requires data at the individual level. In the UK, the English Longitudinal Survey of Ageing (ELSA) reports on activity and ageing.[10] The ELSA is a resource of information on the health and social wellbeing and economic circumstances of the English population who are aged 50 and older. ELSA's current sample contains data covering 14 years. It is similar to the long-running longitudinal survey in the United States. The University of Michigan's Health and Retirement Study (HRS) is a longitudinal panel study that surveys a representative sample of approximately 20,000 people in America, supported by the National Institute on Aging and the Social Security Administration.[11] Through its in-depth interviews, the HRS provides multi-disciplinary data for researchers to use and address

---

10 https://www.elsa-project.ac.uk/about-ELSA
11 https://hrs.isr.umich.edu/about

questions about the challenges and opportunities of ageing. There are a number of other data collection efforts across the world. The ELSA website provides the following resources.[12]

- Brazilian Longitudinal Study on Ageing and Well-Being
- Canadian Longitudinal Study on Ageing
- Korean Longitudinal Study of Ageing
- Health and Retirement Study
- Australian Longitudinal Study of Ageing
- Survey of Health and Retirement in Europe
- Irish Longitudinal Study of Ageing
- Northern Ireland Cohort Longitudinal Study of Ageing
- Healthy Ageing in Scotland
- Mexican Health and Aging Study
- Gateway to Global Ageing
- New Zealand Health, Work and Retirement Survey

The Gateway to Global Aging Data provides a global information platform on ageing around the world.[13] The Gateway to Global Aging website provides harmonised data for cross-country and longitudinal analysis. It also provides a digital library of survey questions plus a search engine for finding concordance information across surveys and across waves.

One of the key areas of data collection under ELSA is on activity constraint, which becomes more significant with age. Ball et al. (2011) noted that the data on activity constraints are captured in two ways: first, by the perception of the person and, second, by an objective measure of walking speed. The survey broke down activity constraints into a three-fold classification; none, mild and severe activity constraints. The slowest walking speed rises to almost 40% of respondents aged 80–84. A substantial proportion of the elderly still retain a significant ability to be at least partly active across all age ranges, with over 55% of 70–74 year olds having an unimpaired walking speed. However, a significant minority below 75 face activity constraints and over half do by 75. The ELSA findings indicate that perceived activity constraints are worse than the walking speed. This is important, as perception can indeed capture more than just walking aspects. Ball et al. (2011) state that

> this may in part result from the fact that people's perceptions take into account other issues, such as the ability to climb stairs comfortably or the risk of falling, than does a simple walking measure. When looking at whether someone wants to move, perceptions, rather than objective measures, are likely to be the most important drivers of their decisions.

The increasing life expectancy across the world due to medical innovation, better healthcare availability and increasing level of health awareness imply that a lot more people will be living well into their 80s and longer. For example, at the age of 65, life expectancy for men in the UK is about 18 years and for women 20 years. World Health Organisation (WHO) data shows the figures for regions across the world. As is evident in Table 7.1, there is a significant rise in life expectancy at age 60 throughout the world over 2000–2016.

12 https://www.elsa-project.ac.uk/links
13 https://g2aging.org/

154 *Demographic shifts and housing*

Table 7.1 Life expectancy at age 60 across the world

		Life expectancy at age 60 (years)	Life expectancy at age 60 (years)	Healthy life expectancy (HALE) at age 60 (years)	Healthy life expectancy (HALE) at age 60 (years)
WHO region	Year	Male	Female	Male	Female
Africa	2016	15.9	17.3	12	13.1
Africa	2010	15.4	16.7	11.5	12.4
Africa	2000	14.3	15.6	10.5	11.4
Americas	2016	21.1	24.3	16.4	18.7
Americas	2010	20.4	23.6	15.8	18.2
Americas	2000	19.1	22.5	14.8	17.4
South-East Asia	2016	17.2	19.1	12.7	13.9
South-East Asia	2010	16.7	18.4	12.2	13.2
South-East Asia	2000	15.8	17.5	11.4	12.4
Europe	2016	20.2	24.1	15.9	18.7
Europe	2010	19.2	23.2	15.1	18
Europe	2000	17.3	21.6	13.5	16.7
Eastern Mediterranean	2016	17.5	19	13	13.6
Eastern Mediterranean	2010	17.3	18.7	12.7	13.3
Eastern Mediterranean	2000	16.8	18.2	12.1	12.7
Western Pacific	2016	19.5	22.5	15.6	17.6
Western Pacific	2010	19	22.1	15.1	17.3
Western Pacific	2000	17.8	20.9	14.2	16.3
(WHO) Global	2016	19	21.9	14.8	16.8
(WHO) Global	2010	18.4	21.3	14.3	16.2
(WHO) Global	2000	17.2	20.2	13.2	15.4

Source: Based on data from WHO; for a full list, refer to the source. http://apps.who.int/gho/data/node.main.688?lang=en

Over these years, life expectancy had gone up by almost 1.8 years for male and 1.7 years for female. Interestingly, the healthy life expectancy (HALE) at age 60 had also gone up by 1.6 years for male and 1.2 years for female. Table 7.1 also shows a remarkable regional variation in life expectancy across the world. Africa records the lowest number whilst the Americas and Europe lead the table.

Similar to Table 7.1, if we take a closer look at one country, we can also see regional variation in life expectancy within a country. A number of factors may cause such variation starting from food/nutrition to health care availability and climatic conditions, along with economic variables such income and employment. For example, in the UK, life expectancy is quite high compared to many other countries. Table 7.2 shows regional variation in the UK in life expectancy at age 65 years for males and females from 2000–02 to 2010–12. Over this time period, significant changes in life expectancy have been recorded. Across regions, south of the country appears to have higher life expectancy compared to other regions. There is obviously a large gap between male and female life expectancy as per the trend elsewhere in the world. The north east region of the country shows the lowest life expectancy for both males and females.

As the life expectancy increases, the housing needs also increase. In the aggregate, the country needs to house more numbers of people for longer number of years. This puts significant demand side effects and in the face of significant supply constraints, such upward drive to demand gets reflected in price movements. The shelter or housing needs in those later

Table 7.2 Life expectancy at age 65 (years): United Kingdom, 2000–02 to 2010–12

Male	UK	North East	North West	Yorkshire and The Humber	East Midlands	West Midlands	East	London	South East	South West
2000–2002	15.98	15.1	15.3	15.7	16.1	15.8	16.6	16.2	16.8	16.9
2002–2004	16.36	15.5	15.7	16.1	16.4	16.2	17.1	16.6	17.1	17.3
2004–2006	16.98	16.1	16.3	16.7	17.0	16.8	17.6	17.3	17.8	17.9
2006–2008	17.45	16.6	16.8	17.2	17.4	17.3	18.2	17.9	18.4	18.3
2008–2010	17.92	17.1	17.2	17.6	17.9	17.9	18.6	18.4	18.8	18.7
2010–2012	18.38	17.6	17.8	18.0	18.3	18.4	19.1	18.9	19.2	19.1
Female	**UK**	**North East**	**North West**	**Yorkshire and The Humber**	**East Midlands**	**West Midlands**	**East**	**London**	**South East**	**South West**
2000–2002	19.07	18.2	18.4	19.0	19.1	19.1	19.6	19.5	19.8	20.0
2002–2004	19.22	18.3	18.5	19.0	19.2	19.2	19.8	19.6	20.0	20.1
2004–2006	19.74	18.7	19.1	19.5	19.7	19.7	20.3	20.2	20.4	20.7
2006–2008	20.11	19.2	19.3	19.8	20.1	20.1	20.7	20.8	20.9	21.1
2008–2010	20.55	19.6	19.8	20.2	20.6	20.6	21.1	21.3	21.4	21.4
2010–2012	20.90	20.0	20.2	20.5	21.0	21.0	21.5	21.7	21.6	21.7

Source: Based on data from ONS, UK. For a full list, refer to the source. https://www.ons.gov.uk/peoplepopulationandcommunity/birthsdeathsandmarriages/lifeexpectancies/datasets/lifeexpectancyatbirthandatage65bylocalareasinenglandandwalesreferencetable1

years (65+) are quite different in nature and dynamics. Much of the usual housing market assumptions and factors may not quite work in the similar fashion. Care must be exercised in interpreting the numbers and relationships. An excerpt from Ball et al. (2011) captures the changes in dynamics and unique features of this sector quite aptly:

> everyone needs a place to live and most of the elderly are relatively comfortable and content as owners of their homes. They have almost all paid their mortgages off and so do not generally feel as financially burdened by ownership as they may have done earlier in their lives. But circumstances may change and people may want to move. As people age, health and mobility deteriorates and, if married, there is a growing risk of losing your spouse. Nature kindly does not tell us a long time in advance about the details of our mortality, so when thinking about housing, even when quite old, it is a good idea to think long-term.

The study (Ball et al., 2011) outlines the findings of a major piece of the research sponsored by McCarthy and Stone, a major retirement housebuilder in the UK, on housing for the elderly who choose to live in a specific type of accommodation: owner-occupied retirement housing (OORH). OORH is typically purchased on a leasehold basis and comprises apartment blocks with design features and amenities suitable for old age living, e.g. communal facilities and round-the-clock in-house support and management. As the study indicates, people in OORH express very high levels of satisfaction with the lifestyle.

The supply side of this market is heavily constrained with very low levels of supply and lack of development opportunities. Underlying demand is much greater than the supply. Ball and Nanda (2013) estimated demand-supply gap and made projections on the retirement housing needs in the UK. Table 7.3 shows forecasts of household numbers 65 and above by age cohort and housing tenure from the paper. Due to price pressures, a sizeable part of the market can be unaffordable for a large number of retiree households. As the study indicates, the accommodation is not expensive in itself compared with the general market price levels and it can be about 10% cheaper than the average. However, that may still be unaffordable

Table 7.3 Forecasts of household numbers 65 and above by age cohort and housing tenure

Age cohorts		1 person			2 person		
		65–74	75–84	85+	65–74	75–84	85+
Age cohort totals (000)	Owners						
2033	1138.4	1501.5	1096.8	2015.8	876.1	474.7
	Renters						
2033	348.2	426.6	311.6	616.6	248.9	134.9
Household size totals (000)	Owners						
2033	3736.7			3366.6		
	Renters						
2033	1086.4			1000.4		
Grand total (000)	Owners						
2033	7103.3					
	Renters						
3033	2086.8					

Source: Based on data from Ball and Nanda (2013); for a full analysis and other information on the above estimates, refer to the paper.

Demographic shifts and housing 157

for a large number of elderly owner occupiers. Note that the estimates are based on several assumptions, which if changed, can alter the forecasts.

One of the most important aspects of retirement housing is people's ability to draw on the housing equity that have been accumulated over the working life. With rising long-run house prices, many retirees typically have a sizeable housing equity to draw on to finance consumption at old age including purchase of a retirement home. The ability and amount to which someone can draw on the housing equity can have a significant and direct link with living standards, amenities and wellbeing at old age. It is not straightforward though. The motives of bequests and precautionary savings can come together to determine the behaviour. Crawford (2018) documents the extent to which older individuals draw on their housing wealth, using data from the English Longitudinal Study of Ageing from 2002–03 to 2014–15. The key findings from the study are:[14]

- Around 4% of owner occupiers aged 50+ moved over a two-year period. Moving is slightly more likely among owners in their 50s and early 60s than among owners in their 70s. From age 80 onwards, the probability of moving increases rapidly with age, driven by moves into institutions.
- Current trends suggest over 40% of those who are owner-occupiers at age 50 will move before death. Cumulating probabilities of moving across older ages suggests that over a third of owner-occupiers at age 50 would move by age 70 if they lived to that age, and over half would move by age 90.
- Fewer than 10% of moves were reported to be financially motivated. The most common reasons for moving were to move to a more suitable home, to move closer to family and friends, health-related (particularly among those aged 80 and over) and to move to a better area (particularly among those aged between 50 and 69).
- It is common for housing wealth to be released even when not explicitly moving for financial reasons. There are two ways movers can release wealth: moving out of owner-occupation and moving to a cheaper property ('downvaluing').
- Changes in household composition are strongly related to releasing housing wealth. Unsurprisingly, those who become separated, divorced or widowed are much more likely to move, move out of owner-occupation, and release wealth than those who remain married or single.
- The financial situation of the household is strongly related to the use of housing wealth. Individuals are more likely to move if they report not having enough money to do things. Among those who move, the average amount and average proportion of wealth released are decreasing with financial wealth, increasing with housing wealth, and increasing in the housing wealth to income ratio.

As is evident from the findings from Crawford (2018), there are many aspects of demand for retirement housing that do not conform to the standard housing demand. Moreover, the earlier discussion from Ball and Nanda (2013) shows very significant demand-supply gaps. Policy measures are needed to tackle such demand-supply mismatch. Public policies need to focus on breaking the supply constraints such as land unavailability and high building costs, land use restrictions, incentives to the development community to view this as income-generating

---

14 Crawford, R. (2018) The use of housing wealth at older ages. IFS Briefing Note BN239. The Institute for Fiscal Studies. https://www.ifs.org.uk/uploads/publications/bns/BN239.pdf

158 *Demographic shifts and housing*

development types. However, it is also important to understand the tastes and preferences of the elderly customers, which differs very significantly with the general market preferences. In order to understand the tastes and preferences of retirement housing, Ball et al. (2011) conducted a survey study with 345 residents at 44 recently built McCarthy and Stone developments spread in Britain. The study found several aspects of perceived benefits of the new home such as:

- Location
- Proximity to relatives and friends
- Ease of living in accommodation
- Sociability and security
- Having a house manager present
- The local environment
- Good health facilities
- Ease of travelling elsewhere

Figure 7.17 shows that 23% of the respondents miss the amount of space they had in their previous home and 18% miss their garden along with 14% miss their friend and families. This shows the preference patterns. The number of rooms per person vary significantly across the countries. Table 7.4 shows the average number of rooms across several European countries. Higher-income countries tend to have almost 1.8–1.9 or close to 2 rooms per person (e.g. UK, Germany, Spain, Netherlands, etc.). Other lower-income countries such as Bulgaria, Poland, Croatia, Slovakia, Romania, Hungary, etc. appear to have almost 1 room per person. The

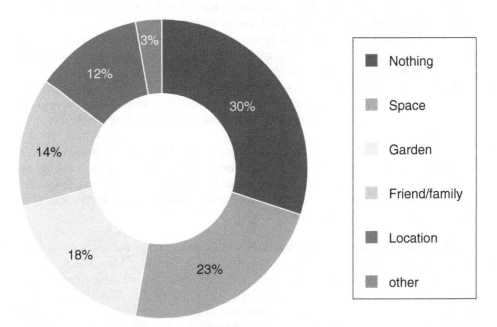

Figure 7.17 Aspect of former homes that are attractive.

Source: Based on data from Ball, M., Blanchette, R., Nanda, A., Wyatt, P., Yeh, J., 2011. "Housing Markets and Independence in Old Age: Expanding the Opportunities".

Table 7.4 Average number of rooms per person in Europe

	2010	2016	2017
European Union (current composition)	1.6	1.6	:
European Union (before the accession of Croatia)	1.6	1.7	:
Euro area (19 countries)	1.7	1.7	:
Belgium	2.1	2.2	2.1
Bulgaria	1.1	1.2	1.2
Czech Republic	1.4	1.5	1.5
Denmark	1.9	1.9	1.9
Germany (until 1990 former territory of the FRG)	1.8	1.8	:
Estonia	1.2	1.6	1.7
Ireland	2.1	2.1	:
Greece	1.2	1.2	1.2
Spain	1.9	1.9	1.9
France	1.8	1.8	:
Croatia	1.1	1.1	:
Italy	1.4	1.4	:
Cyprus	2.0	2.0	:
Latvia	1.0	1.2	1.2
Lithuania	1.1	1.5	:
Luxembourg	1.9	1.9	:
Hungary	1.0	1.2	1.2
Malta	2.0	2.1	2.2
Netherlands	2.0	1.9	2.0
Austria	1.7	1.6	1.6
Poland	1.0	1.1	1.1
Portugal	1.4	1.7	1.7
Romania	1.0	1.0	1.1
Slovenia	1.1	1.5	1.5
Slovakia	1.1	1.1	:
Finland	1.9	1.9	1.9
Sweden	1.7	1.7	:
United Kingdom	1.8	1.9	:
Iceland	1.6	1.6	:
Norway	2.0	2.1	2.1
Switzerland	1.8	1.9	:
Former Yugoslav Republic of Macedonia, the	0.9	1.0	:
Serbia	:	0.9	:

Source: Based on data from Eurostat; for a full list, refer to the source. http://ec.europa.eu/eurostat/web/products-datasets/product?code=ilc_lvho04

problem of under-consumption can be much bigger with the ageing population as they hold on to larger houses with 'empty nester' situations.

Table 7.5 further shows the severity of overcrowding across the income distribution.

Figure 7.18 complements these findings with a question on the reason for moving to a smaller accommodation. It is interesting that as much as a quarter felt no need for space and almost a third felt that the house was too big to look after. This clearly indicates that with suitable retirement housing opportunities, such under-consumption of the housing can be tackled better and the elderly householders are looking for such suitable accommodation. In terms of location, there is a clear preference for staying close to the family members – almost 20% of the respondents.

160 *Demographic shifts and housing*

Table 7.5 Overcrowding rates in households across the income distribution, (2014 or latest year available; Share of overcrowded households, by quintiles of the income distribution, in per cent)

	Bottom quintile	3rd quintile	Top quintile	Total
Poland	47.09%	34.68%	25.36%	35.36%
Mexico	45.43%	37.12%	12.38%	33.27%
Hungary	44.21%	32.23%	24.21%	29.60%
Romania	43.14%	38.19%	33.53%	37.03%
Bulgaria	35.21%	33.66%	31.36%	31.98%
Slovak Republic	32.61%	25.36%	27.82%	28.06%
Sweden	29.71%	7.91%	2.18%	12.59%
Greece	27.71%	16.44%	12.38%	18.66%
Croatia	27.66%	28.37%	30.34%	30.27%
Italy	26.78%	18.24%	11.79%	18.58%
Austria	25.27%	8.20%	4.27%	12.69%
Czech Republic	23.75%	14.18%	9.36%	15.20%
Finland	19.81%	6.49%	1.05%	9.26%
Denmark	16.97%	4.15%	1.56%	7.43%
Slovenia	16.65%	11.68%	5.18%	11.70%
Germany	14.79%	3.82%	1.64%	6.15%
Norway	13.73%	1.78%	0.46%	5.44%
France	13.71%	5.08%	2.11%	6.32%
Chile	11.04%	10.85%	4.13%	9.25%
Netherlands	9.52%	2.29%	0.66%	3.61%
Portugal	9.33%	5.12%	2.41%	5.52%
United Kingdom	8.47%	6.38%	1.94%	5.45%
United States	7.37%	3.24%	1.22%	3.78%
Switzerland	7.31%	4.98%	1.23%	5.28%
Belgium	5.88%	0.57%	0.13%	1.84%
Spain	5.85%	2.02%	1.63%	2.95%
Korea	3.63%	3.63%	1.06%	5.75%
Ireland	3.30%	1.40%	0.96%	2.47%
Japan	1.77%	0.73%	2.53%	1.60%

Source: Based on OECD calculations based on European Survey on Income and Living Conditions (EU SILC) 2014 except Germany; the Household, Income and Labour Dynamics Survey (HILDA) for Australia (2014); Encuesta de Caracterización Socioeconómica Nacional (CASEN) for Chile (2013); the German Socioeconomic Panel (GSOEP) for Germany (2014); the Korean Housing Survey (2015); Japan Household Panel Study (JHPS) for Japan (2014); Encuesta Nacional de Ingresos y Gastos de los Hogares (ENIGH) for Mexico (2014); American Community Survey (ACS) for the United States (2014). For a full list, refer to the source.

There can be several significant benefits of owner-occupied retirement housing as noted by Ball et al. (2011).

- **Higher quality of life for its residents**. The report notes that 92% of elderly residents are very happy or contented and most would recommend it to others.
- **Improved health for residents and reduced impact on the NHS**. As the accommodation is designed for dealing with activity constraints, residents are able to manage their health conditions better and may possibly require less emergency health care and call on less frequently for hospital services and stays. Health services across the world report high cost of in-patient care and any factor that lowers the probability of in-patient care is highly cost-saving.
- **Good for the environment**. 51% of the surveyed residents said that their energy bills reduced significantly and was less than what they had to pay in their previous

Demographic shifts and housing 161

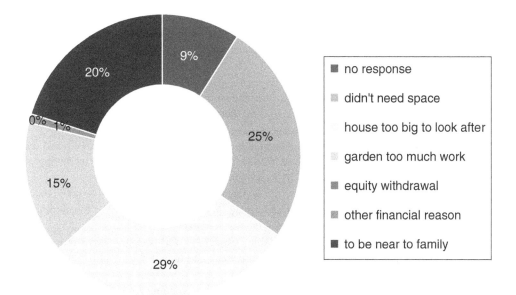

Figure 7.18 Reason for moving to smaller accommodation.

Source: Based on data from Ball, M., Blanchette, R., Nanda, A., Wyatt, P., Yeh, J., 2011. "Housing Markets and Independence in Old Age: Expanding the Opportunities".

homes. Part of the reason is the newly-built and purpose-built retirement facilities show much better energy efficiency levels compared to older homes or the general market housing stock. Also, since people tend to travel less in retirement as being close to friends and relatives and to shops and other amenities, it reduced the per-capita carbon footprint. An interesting spillover effect of moving into retirement housing is that it also allows the new owner of the previous home (of the retiree) to undertake renovations to improve the energy efficiency with positive environmental benefits.

- **Boosting local neighbourhoods.** Older people in retirement facilities have significant purchasing power which manifests in more revenues for local shops. As the study notes, many local market-based services face a threat with the growth of out-of-town shopping and online shopping, but the elderly people prefer to use local amenities more than many other residents.
- **Positive impact on local housing markets**. Typically, elderly households occupy a sizeable home with significant equity. Under-consumption is very common as children have left the house. Therefore, availability of suitable retirement housing can enable recycling of those houses to bigger families who need them and at the same time providing personal wellbeing-improving option for the elderly householder. This turnover is desirable for a healthy housing market. Table 6.2 in chapter 6 shows more than 30% of housing stock is owned outright in England and Wales, according to the 2011 census. This implies that a sizeable number of these houses can be held by the elderly households who have paid off their mortgages, which is typically achieved at the retirement stage.

Devising policies suitable for tackling retirement housing is challenging due to several reasons. There is typically lack of understanding of the level of demand and supply at the granular level. This can be true in any country. Local planning policies may be silent on this subject. There is a much greater need to recognise the problem and have specific strategic measures in place. Policies are required for:

- Release of developable land specifically for retirement housing.
- Inclusion of retirement housing in general market development to create mixed and balanced communities. This is particularly favoured by the elderly residents.
- Devise engagement and employment schemes for elderly residents. This can boost the utilisation of elderly human capital as well as ensure mental and physical wellbeing of the elderly cohort, with much benefits to society and economy.
- Amend building regulation in a way to allow developers to achieve operational efficiency.
- Recognise the shift in service needs and create an employment and skill development programme suitable for such shifts in the labour market.
- Make retirement housing more affordable, especially for people whose retirement income streams are not strong and benefits are weak and do not have significant housing equity.
- Create awareness programme for appropriate financial planning for retirement.
- Promote healthy ageing awareness.

In this chapter, I have discussed the effects and implications of demographic changes such as the ageing population and migration in the housing market. One of the significant constraints that migrants and young people (and other households) face is access to the mortgage market. The housing finance system has evolved significantly since the 1980s with much complexities in place now. In the next chapter I discuss the housing finance market and associated developments.

Selected research topics

- Housing policies suitable for integrating migrant population.
- Future tenure choices for young people.
- Linkages between retirement planning and housing.
- Association between cognitive processing, individual preferences, and personal economic and living outcomes at old age.
- Effect of ageing population on economic and social dynamics within and between spatial territories.

8 Housing finance and the mortgage market

Chapter outline

- Structure and fundamentals of housing finance
- Mortgage market dynamics
- Lending channels
- International comparison of housing finance systems
- Selected research topics

Since the 1980s, much development in financial and mortgage markets have happened and significantly changed housing market dynamics across the world. Housing finance schemes have become very common ways of funding house purchases. Private lenders as well as public lenders have emerged in a big way to enable buying houses. Personal finance and credit market dynamics come into play in many complicated schemes and structures. While much of the developments in the mortgage market and associated products and services happened in the developed countries, especially in the United States of America, much of the rest of the world have followed the practices. The widespread use of mortgage finance has shaped a significant part of the economic policies around the world.

For households, a mortgage loan is perhaps the most significant and sizeable financial commitment and liability. Since a house is a very important and frequently used engine of creating personal wealth for households, it is a critical aspect of the household's financial wellbeing. A careful management of the housing finance and prudent decisions at important junctures of the loan life cycle is important for stability and improvement in personal wealth through housing. As discussed in the section on retirement housing in the previous chapter, reliance on housing wealth or equity has become an important way of creating cash flow and fund consumptions at old age. Therefore, the right decisions made on mortgage financing can be an important factor for ensuring such 'old age insurance'.

The importance and structure of mortgage finance has also become very important for the economic wellbeing of the countries. It is critical for ensuring that transmission channels of monetary policy measures work in a desired manner. Successful management of the housing finance sector is essential for effectiveness of monetary policy instruments. This is due to the fact that much of the financial service sector across the world is tightly connected to the monetary policy dynamics. However, we have also seen repeated disruptions to these relationships and breaks in the dynamics of this nexus have resulted in several financial and banking crises, for example, the 2008 Global Financial Crisis and crises in Ireland and Spain. In an interesting finding, Claessens et al. (2011) noted that when such crises are linked to housing cycles of boom and bust, those tend to be more severe and devastating than other economic crises. This shows the depth and breadth of the relationship of the housing finance sector and financial sector. As discussed before in earlier chapter, housing is a unique economic good with both consumption and investment motivations. Housing is a good with significant connection with family and society. A good, stable and healthy housing market can promote much progress for the economy as well as society with less frequent loss of equity or financial fortune. In the past economic and financial crises, many families and households have suffered due to loss of home equity and foreclosures. Therefore, the housing finance issues should be analysed from how those can affect and have significant implications for socio-economic development perspectives. As the world population is becoming more and more urbanised, the needs of housing is becoming greater and more complex. As more and more governments are undertaking massive and complex development projects to provide housing for the increasing population with ever expanding aspirations for having their own homes, it is very crucial to have a well-functioning housing finance system that can promote financial as well as socio-economic stability. A vast section of the population across most developing economies of the world are progressing towards middle-class category and with more purchasing power, they are aspiring for higher level of consumption including owner-occupied housing. Countries such as India, with a massive burgeoning middle-class population, and China, with huge urbanisation and a population with a rising purchasing power, now require huge numbers of houses and the housing finance system in those countries are starting to get more formal and regulated. Since mortgage financing is long-term (usual 25–30 years of tenure), in countries with short-term or informal financial contracts, development of mortgage financing can enable strengthening and lengthening of financial contracts. Beck et al., (2011) argue that a well-functioning mortgage finance system can be effective in developing the non-banking financial sector. However, it is important to note that a well-functioning mortgage finance system can work if there are tight regulations and controls for credit practices by the consumers and lending institutions.

There is a huge variation around the world in terms of mortgage market development and use of mortgages to finance housing. Some countries, mostly developed economies, have a high percentage of housing purchases financed by mortgages. A World Bank report in 2014 analysed new data on the depth and penetration of mortgage markets across countries and found a large variation across both dimensions of mortgage market development, across countries, but also – in terms of depth – within countries.[1] The authors note that mortgage markets appear to develop only at relatively high levels of gross domestic product per capita. This is no surprise, as purchasing ability of the consumer fuels the market activities and the development. With the development comes the need for policies associated with the financial system. Price stability, healthy functioning of contractual frameworks, efficiency of the information flow, etc. need to be maintained or restored by policy prescriptions. Due to the complexity of the mortgage market and its relationships with other parts of the financial system, such as the insurance sector and the stock market, which are sources of long-term funding, mortgage market development critically depends on the development of those parts of the financial sector.

In most developed and high-income countries, many people avail of mortgage finance system to fund their home purchases. However, the World Bank Report (2014) finds that many low and lower-middle-income countries only have a tiny share of the population availing of mortgages. The study looked at this from two metrics – housing loan penetration and mortgage depth. Housing loan penetration is defined as the percentage of the adult population

---

1 Badev, A., Beck, T., Vado, L., Walley, S. (2014) Housing finance across countries: New data and analysis. Policy Research working paper; no. WPS 6756. Washington, DC: World Bank Group. http://documents.worldbank.org/curated/en/697351468165251669/Housing-finance-across-countries-new-data-and-analysis

with an outstanding loan to purchase a home, which was obtained from the Global Financial Inclusion (FINDEX) Database.

Mortgage depth is defined as the outstanding loan to GDP ratio. Table 8.1 shows the housing loan penetration for several countries. It lists the countries with high and low extents of mortgage use. It also reports numbers for BRICS countries. There is a huge variation. Sweden boasts of almost 60% of the adult population having accessed the mortgage market and having the loan obligations. Many major developed countries such as the US have more than a third of the population having mortgage obligations. The UK, Spain and Canada show similar penetration rates. Several African countries exhibit very low (almost 1% or less) penetration (on the right side of the table). Mortgage markets in those need much more development. Much work is needed to improve the penetration in those countries. More often we find the presence of a sizeable informal sector in those countries with a low penetration of mortgage finance system. Lack of regulation and governance standards, absence of major banks and lending institutions, political instability and above all, very low per capita income can be the major factors behind such a low rate of penetration.

In the middle of the table, it shows the penetration rates for two most populous countries (India and China) in the world along with the other members of the BRICS countries whose combined GDP is very significant and likely to be the most dominant economies in the world in the future. However, a very small portion of the population in such populous countries have accessed mortgage financing – for example, China has just above 5% and India has only 2.5%. Much more activities in the financial sector are being undertaken. India, since the

Table 8.1 Housing loan penetration by country (2011)

With high housing loan penetration	Score	BRICS	Score	With low housing loan penetration	Score
Sweden	59.7	China	5.4	Niger	1
Denmark	53.5	South Africa	5.4	Turkmenistan	1
Netherlands	45	India	2.5	Indonesia	0.8
Australia	43.8	Russia	1.8	Lesotho	0.8
New Zealand	42.6	Brazil	1.6	Mali	0.8
Luxembourg	41.5			Burkina Faso	0.7
Belgium	39.2			Comoros	0.7
Ireland	39.2			Moldova	0.7
US	36.7			Benin	0.6
Spain	35.3			Gabon	0.6
UK	34.3			Kyrgyz Republic	0.6
Canada	33.4			Madagascar	0.6
Finland	33.1			Tajikistan	0.6
France	31.1			Venezuela	0.6
Cyprus	29.2			Argentina	0.5
Austria	29.1			Georgia	0.5
Portugal	25.9			Sierra Leone	0.5
South Korea	25.6			Azerbaijan	0.4
Germany	23.7			Nicaragua	0.4
Kuwait	23.5			Congo	0.3
UAE	23			Nigeria	0.3
Singapore	21.8			Guinea	0.2
Malta	20.5			Burundi	0.1
Qatar	20.4			Senegal	0.1
Estonia	19			Uzbekistan	0

Source: Based on data from Badev, Beck, Vado and Walley (2014); for a full list, refer to the source.

Table 8.2 Total outstanding residential loans to GDP ratio, 2004–2014

	2004	2006	2008	2012	2014
Austria	19.9	23.2	24.7	27.2	27.5
Belgium	30.0	34.9	38.6	47.3	49.1
Croatia	8.5	12.9	15.6	18.9	18.3
Czech Republic	4.0	6.7	9.4	13.5	16.6
Denmark	85.6	98.7	106.9	114.1	114.0
France	25.3	30.9	35.1	41.7	43.3
Germany	51.0	49.5	44.7	43.0	42.4
Greece	17.6	26.2	32.1	38.4	38.8
Hungary	9.3	15.1	20.9	20.2	16.6
Ireland	49.9	67.5	79.6	57.1	49.4
Italy	12.8	15.8	16.2	22.6	22.2
Latvia	11.2	26.9	29.5	24.0	19.5
Lithuania	6.9	12.4	18.5	17.4	16.4
Luxembourg	33.7	36.0	42.3	49.8	50.7
Netherlands	82.7	88.3	92.6	101.2	95.7
Poland	4.7	8.3	14.2	20.6	20.0
Portugal	46.7	55.3	58.8	65.6	59.2
Romania		2.2	4.0	6.6	6.7
Slovak Republic	6.3	11.5	13.0	19.0	23.1
Slovenia	2.9	6.2	9.0	14.6	14.4
Spain	44.7	56.7	60.4	60.8	55.4
Sweden	53.3	61.3	58.5	79.1	78.8
United Kingdom	67.3	78.0	67.5	76.1	75.0
Norway	53.1	55.0	49.7	66.3	75.8
Turkey	0.1	1.5	2.6	5.9	6.5
Russian Federation		0.6	1.9	3.6	3.5
United States	49.0	57.8	72.9	75.0	67.6

Empty cells: Data not available.
Source: Based on data from European Mortgage Federation (EMF), Hypostat (2015), A Review of Europe's Mortgage and Housing Markets, p. 90. For a full list, refer to the source. https://www.cesifo-group.de/de/ifoHome/facts/DICE.html

early 1990s, has taken major steps in liberalising the economy and financial sector and has established a robust regulatory framework for the financial sector.

Table 8.2 shows a different way of looking at how important and big the mortgage market activities across the OECD countries is. Denmark has a remarkable pattern with more than 100% of the GDP as the residential loan outstanding. The UK has almost 75% mortgage depth, while the US shows about 67%. This can be very advantageous for the economy's performance as accumulation of home equity can fuel further consumption due to housing wealth effect, which would further boost aggregate spending. However, weakness in the system and having bad loans can be devastating due to significant exposure to the mortgage finance. Moreover, it is also crucial to limit or manage the exposure to global capitals in a way that does not expose the domestic economy to the frailties of the global market.

Mortgage market process

The journey of a mortgage customer is quite similar across the countries. It typically starts with a house search and once the housing search is completed and an offer is accepted, the mortgage financing journey commences. The first step of the mortgage financing journey is to call on the bank (typically a local branch of a bank) and request a mortgage. Banks, depending on meeting the affordability criteria and desirable credit worthiness, may agree to finance the house purchase. With submission of the mortgage deposit (typically 10–20% of the house value), the exchange of contracts takes place and the transaction is completed. From then on the householder agrees to pay and pays monthly mortgage obligation payments.

Mortgage enables prospective homeowners to afford to purchase a house by paying the loan obligation with small portions each month. With a well-justified promise and prospect of a steady stream of income over the mortgage life cycle, an individual can borrow the large sum of money needed to finance a house purchase. He/she obviously has to pay interest for the loan amount, which depends on the market condition, benchmark interest rate and credit worthiness of the borrower. Here is an example of how a mortgage would work out for an individual household:

Suppose a household (the borrower) wants to purchase a house priced at £220,000. The household can pay a maximum £20,000 towards the deposit. This means he/she needs to borrow £200,000. Suppose the householder is offered a 30-year, fixed-rate mortgage with a contract interest rate of 6% annually by the bank. The question is how much will this borrower have to pay each month to pay off the loan in 30 years?

The formula to use for the calculation of monthly loan obligation is:

$$MP = MB_0 (1+r)^n \frac{r}{(1+r)^n - 1} \tag{8.1}$$

Where MMP is the monthly mortgage payment, IMB_0 is the initial mortgage balance which is the amount borrowed. n is the number of months over which the loan will be amortised, and r is the monthly interest rate (annual interest rate divided by 12 months). Consequently, the monthly payment is calculated using the equation (8.1):

$$MP = £200,000 \{1+(0.06/12)\}^{360} \frac{0.06/12}{\{1+(0.06/12)\}^{360} - 1} = £1,199.10 \tag{8.2}$$

So, the householder will need to make a commitment of paying almost £1,199 per month for 30 years. At the end of the 30 years loan tenure, he/she will have the full 100% equity of the house. Not only will the homeowner have the monthly loan obligation of £1,199.10, he/she will also have other fees and services as the cost of homeownership. Some of the fees are linked with the mortgage. It is fairly common for homeowners to buy insurance products and services in the amount of the loan amount to cover for the events that might restrict or hamper the homeowner's ability to pay the monthly instalments and steady stream of income. Therefore, the homeowner will have to pay a mortgage insurance premium. Moreover, the homeowner will need to purchase home insurance to cover for the physical damage to the house. Other related costs of homeownership remain the same such as council or property tax, maintenance, etc. Some lenders also charge one-time fees for the mortgage application, e.g. £1,000, which is often included in the monthly payment, although that is optional. The mortgage application process is not straightforward. In many countries, consumers often go through mortgage advisors to be able to choose from numerous mortgage products; while the basic features of those are quite similar, they do differ in several ways.

Types of mortgage financing

Mortgages vary significantly in terms of the following features:

- Interest rates: This is a key consideration and differentiation. Even 10 basis points difference in the interest rate can make a significant difference in the amount of monthly payment. Therefore, consumers search hard on getting the cheapest possible rate. However, the cheapest possible mortgage rate may not always be the best option, as some other features may be more important from the consumer's particular perspectives. Interest rate often is the most important factor behind the decision to refinance or re-mortgage.
- Mortgage tenure: While typically people take out mortgage loans for 25–30 years, there are mortgages with shorter and longer tenures. Depending on the income level, employment type, retirement plan and wealth, homeowners choose various tenures.
- Various charges to pay: mortgages do differ in terms of various fees that need to be paid at various stages of the mortgage life cycle such as application fee and pre-payment fees.
- Repayment scheme: Most mortgages are straightforward repayment of the principal amount along with an interest component. But, there are also mortgages which are interest-only loans, where the borrower needs to pay just the interest component each month and will repay the capital/principal at the end of the mortgage tenure, which can come from the savings over the mortgage life cycle.
- Mortgages of specific types: There are mortgage products specially designed for first-time buyers or landlords who would like to buy houses for renting out (buy-to-let mortgage in the UK), shared-ownership mortgages in the UK, or mortgages for the elderly customers, which are often called reverse mortgages. A reverse mortgage is also a loan for homeowners who are old (e.g. 62+) and have significant home equity, which enables them to borrow against the value of their homes and receive a lump sum, fixed monthly payment or a line of credit. A reverse mortgage comes with no monthly loan payment. The loan balance is payable when the borrower dies or moves out. There is one tricky issue when the market value of the home drops or the borrower lives on for a long time. There are regulations that requires lenders to structure the loan with the loan amount not surpassing the home's value and there is no recourse if the loan balance is larger than the home's value.

The main types of mortgages are as follows:

- **Annuity mortgages**
  This is the most common types of mortgages, where the principal or the capital is paid off along with interest charges over a long time period. The interest rates vary a lot along with various scenarios, as explained below.
- **Balloon mortgages**
  The whole principal amount is not paid off during the term. Whatever is not paid will have to be paid after a given (often relatively short) time period. This can be done through refinancing the mortgage.

There are two main types of risks in mortgage financing from the lenders' perspective.

- **Prepayment risks and associated penalties.**
- **Default risks and associated foreclosure.**
- These affect the flow of returns to investors and the extent to which mortgage securities have bond-like characteristics.

There are many types of mortgages in the market with significantly different products and more nuanced offerings. Some of them are as follows:

1 **Repayment mortgages**

   For the repayment mortgages, the borrower repays a combination of principal and interest components each month. The monthly payment, as shown in equation (8.1), is computed in a way that the outstanding principal amount becomes zero at the end of the mortgage tenure. The borrower then becomes the full owner of the house with 100% equity. When people move during one mortgage tenure, they often take another mortgage or carry the same mortgage to the new home. However, more often, as the house value goes up over time, it is likely that the homeowner might be able to pay off more than as computed at the beginning. Refinancing is quite common, as due to better financial health and credit worthiness, many homeowners may be able to get better credit terms on the new mortgage. There are mortgages which allows consumers to pay off flexibly (called flexible mortgages) depending on their financial and life conditions, although such mortgages can be more expensive in terms of interest rates.

   - **Interest-only mortgages**

   With the interest-only loans, the borrower pays just the interest component each month and repays the principal or the capital at the end of the mortgage tenure. Either savings or inheritance or sales proceeds need to be arranged to pay off the capital. Lenders would need to be satisfied with a repayment plan of the principal. It might be possible to switch to repayment mortgage if the borrower wishes to do so. The most important and attractive feature of the interest-only mortgage is that the monthly payment is much lower than any other mortgage products.

   - **Fixed rate mortgages**

   There are mortgage products for which the interest rate is fixed for a certain number of years such as 2, 3 or 5 years. During the initial 2, 3 or 5 years, the interest rate stays the same as at the initiation. In the rising interest rate environment, this is of significant value to the buyers. This is also useful for financial planning as consumers are more confident about what they need to pay each month for a certain amount of years. At the end of the fixed-rate period, the interest rate is variable, which is often higher than the fixed rate. Sometimes consumers take up another fixed rate mortgage at the end of the current mortgage fixed-rate period.

   - **Variable rate mortgages**

   Compared to the fixed rate mortgages, if the expectation is that the interest rate may fall in the short-run or in the near future, consumers can opt for variable rate mortgages, for which the rate can increase or decrease over time. Some lenders can offer a discount on the standard variable rate for a certain number of years, about 2–5 years, which are called discount rate mortgages.

   - **Cashback mortgages**

   There are lenders who sometimes offer a cashback (a percentage of the loan) for the mortgage issued. However, the rate can be higher than other mortgage products. While it can be attractive for someone looking for cash, in the end it may not work out as the cheapest mortgage overall.

- **Offset mortgages**

This is the type of mortgage where the interest is calculated on the difference of the principal amount and the balance on the linked savings account. For example, if a consumer has a mortgage principal amount of £200,000 and also has a linked savings account with £20,000, then he/she only needs to pay interest on the difference, i.e. £180,000. It is possible for the consumer to access the savings account when needed but they would not earn any interest. Moreover, the rate on the offset mortgages tend to be higher than other available products.

- **First time buyer mortgages**

Across many countries, in order to boost homeownership, lending institutions are often backed by government support and subsidies (e.g. help-to-buy mortgages in the UK) and offer more favourable mortgage products to first-time buyers. This can be in the form of allowing higher than usual loan-to-value ratio or a cheaper interest rate or a combination of the two.

- **Buy to let mortgages**

In order to boost the private rented sector, there are mortgage products that are specifically designed for landlords to purchase a house in order to rent it out. These are called buy-to-let mortgages. Perceived rental income often come into play in the calculations for the mortgage.

Figure 8.1 shows a process map of the mortgage market mechanism. It demonstrates how a mortgage is originated or initiated in a local market, which is triggered by a need and subsequent demand by an individual household. Once the loan application is accepted and a mortgage is issued by the lending institution, this is the primary mortgage market. The secondary mortgage market almost kicks in depending on what lending institutions do with the mortgage issued. They can hold it in their own portfolio, which depends on the type and strength of the lending institution. They can also sell the loan in the secondary market, from which many types of investors such as pension funds, life insurance companies,

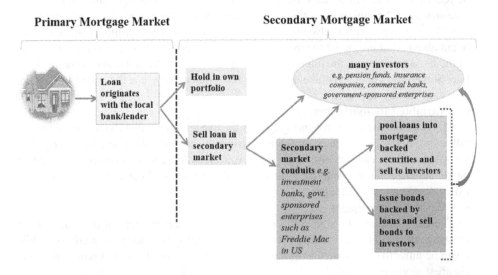

Figure 8.1 Mortgage market process.

Source: Adapted from DiPasquale and Wheaton, 1996, Urban Economics and Real Estate Markets, Prentice-Hall.

commercial banks and in the case of the US, government-sponsored enterprises such as Fannie Mae and Freddie Mac.

The Federal National Mortgage Association, commonly known as Fannie Mae, is a United States government-sponsored enterprise. It is a publicly traded company. The Federal Home Loan Mortgage Corporation (FHLMC) is known as Freddie Mac. It was created in 1970 to boost the secondary mortgage market functioning in the US. Along with the Federal National Mortgage Association (Fannie Mae), Freddie Mac is tasked with buying mortgages on the secondary market, pooling them into mortgage-backed securities, which are then sold to other investors in the open secondary market. A mortgage-backed security (MBS) is a type of asset-backed security that is secured by a mortgage or a number of mortgages (explained later). In the secondary market, investors can buy and sell the MBS. The secondary market conduits such as investment banks and government-sponsored enterprises can also issue bonds, which are backed by mortgage loans, and sell the bonds to the investors. The buying and selling activities among the investors keep the secondary market live and functioning. However, the role of credit agencies is crucial in terms of rating the loans appropriately based on risk criteria. The importance of the secondary mortgage market is high as it can inject more money available for mortgage lending, which would enable households to afford new home purchases. A healthy and well-functioning secondary mortgage market can keep the housing market accessible for a lot more households and enable them to realise their home-ownership dreams.

Moreover, the intermediation aspect of the mortgage providers may have a role in improving overall social welfare by matching the savings objectives of retail and capital market savers with consumption and investment motivation of mortgage borrowers. Mortgage financing has the potential to increase the efficiency of the housing market by raising the scale and scope of housing transactions. However, this can also lead to the possibility of asset-liability and maturity mismatch due to the fact that a bank could also have substantial long-term assets (such as fixed-rate mortgages) funded by short-term liabilities, such as retail deposits. If short-term interest rates rise, the short-term liabilities would be up for re-pricing at maturity, while the yield on the fixed-rate assets, which are longer-term, may remain the same. This implies that while the income sources may not have changed, the costs of assets have increased, leading to what is known as a maturity mismatch. In a significant work using historical data, Jordà et al. (2016) established that a high level of maturity mismatch in the banking sector could step from their historically largest expansion in mortgage lending.[2] Another important finding of the study is that mortgage credit has played a major role behind the financial fragility across the developed economies. Authors also presented evidence that mortgage lending booms were only weakly linked with financial crisis risks before World War II, but since then real estate credit has become a significant predictor of the financial fragility across the countries. Derailment of the mortgage market seemed to have heavily shaped the nature and extent of business cycle booms and troughs.

Secondary mortgage market products

A mortgage-backed security (MBS) is a type of asset-backed security, but in this case, it is secured by a mortgage or a collection of mortgages. The MBS need to be classified in

---

2 "The Great Mortgaging: Housing Finance, Crises and Business Cycles" 2016 (Jordà, O., Schularick, M., Taylor, A.M.) *Economic Policy* 31: 85 107–152. https://doi.org/10.1093/epolic/eiv017

rating categories as determined by the credit rating agency. A MBS holder receives periodic payments much like the coupon payments. There are regulations that restrict the origination of the mortgages, however, laxity in those standards can also creep into the system, as we have seen prior to the Global Financial Crisis of 2008. In many ways, banks or lending institutions in the primary market acts as a conduit or agent between the homeowner and the investors in the secondary market. It allows the banks to have sufficient finances to be able to lend beyond the reliance on their retail savings. The role of the investors is indirect, but very much intertwined into the housing market as MBS can also be used for the interest and principal payments to the stakeholders of the mortgage pool (as per Figure 8.1). However, the payment depends on the riskiness of different mortgages as they are classified under the MBS.

Two common types of MBSs are offered in the market – pass-through and collateralised mortgage obligations, also known as CMOs. Many types of mortgages can be combined to create the pass-through, such as the adjustable-rate mortgages (ARM), fixed-rate mortgages and other types of loans. Pass-through MBSs work under a trust in which mortgage payments are deposited and then 'passed through' to investors. Typically, 5, 15 or 30 years of maturities are allowed under pass-through mortgage-backed securities, although, in reality, it may last less than the maturity as it depends on the mortgage lives. On the other hand, collateralised mortgage obligations (CMO) are more complex, and works more like a collateralised debt obligation (CDO) but comprises only MBSs. These are complicated as those tend to have several pools of securities, which are then 'sliced or diced' into tranches of specific credit ratings. The ratings determine the interest rates to the investors. The usual risk-return trade-off comes into play with the senior secured tranche having lower interest rates than the unsecured tranche, as the higher risk needs to be compensated with higher interest rates. CDOs and CMOs have been under the scanner from the global financial crisis in 2008. The inaccurate risk ratings and misplaced pricing can lead to severe losses and with so many investors and a huge volume of investment attached to these instruments, the bust can lead to widespread financial crisis such as the GFC in 2008. Several investment banks and companies went down due to their excessive exposure to the CDOs/CMOs. The crucial factor is the proper evaluation and matching of risks and return, based on sound economics of demand and supply, as discussed in previous chapters. Any laxity in those can be disastrous.

So, the question is: why do secondary markets function? Risk-return appetite varies among investors. They can be classified into risk-taker, risk-neutral or risk-averter. There is large and well-established literature on risk types and patterns of risk aversion such as constant absolute risk aversion (CARA) and constant relative risk aversion (CRRA). The secondary market activity can bring a lot more investors into the housing market actions and create a liquid market that can provide more funds. Secondary market, if functioning effectively with good governance structure, can address issues of moral hazard and adverse selection in the mortgage market. Competition and innovation in products and services can keep the market structure efficient and healthy. However, the key to such scenarios is to ensure a well-regulated system. With the advent of CDOs/CMOs, many investors, globally, can get involved in the secondary market. While it is helpful for injecting investment into the housing market, it can also expose a large section of the investment community. With an integrated global investment market, significant weakness in one market can have cascading or contagion effects, as we have experienced during the GFC in 2008. While safeguarding the mortgage finance system by interventions such as bailing out 'too big to fail' lending institutions can be justified on one ground, it can also inject moral hazard issues into the system. With global integrations, it is hard to monitor the mortgage market unless there is a concerted effort from all corners of the market globally. However, the mortgage market can still provide a better solution for housing as economies with weak or non-existent mortgage markets have experienced more severe housing problems. In the 2008 crisis, one of the main issues was the exposure and dominance of bad loans or subprime loans, which was caused by laxity in lending standards and requirement. Sub-prime lending is about providing loans to households with difficulty obtaining loans from the prime market. This could be due to multiple reasons, most often impaired credit and low incomes are the main reasons. These loans come with high risk and so attract high interest rates, which cause further issue of affordability if the household does not have high and steady income. The default risk and rate run high on these loans. Sub-prime lending may work when the housing market is expanding and prices are rising. Sub-prime borrowers can keep on transacting and go away without defaulting. However, there is always a declining market ahead of a boom, which can make a lot of these loans unaffordable and especially with weak labour market, this can lead to widespread foreclosures. Not just subprime market, but also the prime market players can also be affected if households take out very large ('jumbo') loans and home-equity loans – living mostly on leverages. Ratings agencies have a significant role to play – both in terms of estimating the grades for MBS tranches but also for the borrowers. The risk-return trade off can only work well if the risk is estimated accurately at all levels.

To be able to evaluate and estimate risk, rating agencies require a large amount of granular data. While in the past much of the estimation of risks were based on historical data, with much stronger computing ability and big data analytical methods, it is now possible to inject more data into the estimation process. With richer sets of data items and large datasets, the problem of omitted variable bias or unobserved heterogeneity can be better addressed, which should enable the ratings agencies to predict more accurate risk patterns. However, crucial is still the governance structure and restrictions of predatory financing behaviour. A huge uptick in sub-prime lending activities have been proved to lead to severe risk fallacy with disastrous consequences in the past.

Sources of mortgage financing

Every country has its own residential mortgage system. The institutional histories in context of broader financial systems, regulations and housing policies are important factors in shaping the mortgage financing system in each country. There are several sources of mortgage finance, such as follows:

- Retail savings deposits with the banks and other financial institutions, e.g. banks and in the UK, building societies receive deposits which are used for mortgage financing.
- Covered bond markets, for example, the German Pfandbriefe, which are a type of covered bonds issued by the German mortgage banks, and these are collateralised by long-term assets. These types of bonds represent the largest segment of the German private debt market and are considered to be the safest debt instruments in the private market. Another example is the Danish mortgage bond.
- Mortgage backed securities: pools of mortgages are used as the asset base for structuring residential mortgage-backed securities (RMBS). Often, these are packed into special purpose vehicles such as CMOs as discussed before.

- Inter-bank lending: Short-term money markets (inter-bank lending) fund bank mortgage lending.

Moreover, countries differ in terms of government support for the mortgage financing system. Governments can play several roles in the mortgage market such as an insurer, security guarantor and also as issuer and investors in the secondary market.

Table 8.3, based on information from Lea (2010), shows the variation in terms of government support across major economies. The United States with its most complicated and developed mortgage financing system has all three types of government-supported mortgage institutions or guarantee programmes: mortgage insurance, mortgage guarantees and government-sponsored mortgage enterprises. Canada has government guarantee programmes as well as government-backed insurance programmes. Japan has government guarantee programmes. Netherlands has government-backed mortgage insurance programmes. South Korea has a GSE similar to those in the United States. However, the US dominates in terms of the government's active role in bolstering the secondary mortgage market by creating huge government-sponsored enterprises such as Fannie Mae and Freddie Mac.

Lea (2010) also describes the sources of mortgage financing across a number of major economies with a sizeable housing finance sector in 2006, just prior to the global financial crisis of 2008. The time period captures almost the peak of the mortgage financing industry in that boom period. Banks such as commercial, savings and cooperative banks are a significant source of mortgage financing in most countries and the banking sector carries the largest share of mortgage lending across most countries. The banking institutions significantly drew on their deposit funding in several leading economies – such as Australia, Canada, Japan, Netherlands, Germany, Switzerland and the UK. There are notable exceptions such as Denmark with a very large share of the mortgage bonds; Spain having a significant share of mortgage bonds and the MBS and the US with a sizeable share for the MBS. It is also mentioned in the report that the ARMs offered an attractive option for the banks to hold it in their portfolio as those minimised the interest rate risk. Of the ARM countries in the study,

Table 8.3 Government support to mortgage market

Country	Presence of government mortgage insurer	Presence of government security guarantees	Presence of government sponsored enterprises
Netherlands	NHG	No	No
UK	No	No	No
Australia	No	No	No
Canada	Canada Mortgage and Housing Corporation (CMHC)	Canada Mortgage and Housing Corporation (CMHC)	No
Japan	No	Japan Housing Finance Agency	possible
Korea	No	No	Korean Housing Finance Corp.
US	Federal Housing Administration	Government National Mortgage Association (GNMA or Ginnie Mae)	Fannie Mae, Freddie Mac, Federal Home Loan Banks (FHLBs)

Source: Based on data from Lea, M. (2010) International Comparison of Mortgage Product Offerings. Research Institute for Housing America; for a full list, refer to the source. https://cbaweb.sdsu.edu/assets/files/research/Lea/10122_Research_RIHA_Lea_Report.pdf

Spain seemed to have significant exposure to the capital markets for funding. Almost 70% of funding came from covered bonds and securitisation activities such as MBSs. The rapid growth of mortgage lending and the acceptance of AAA-rated security tranches and covered bonds as repo collateral at the European Central Bank (ECB) might have fuelled the high use of the capital markets. The Lea (2010 report notes that

> funding availability and characteristics are also major factors in the dominance of short- to medium term fixed-rate mortgages in many countries. In developed markets, such instruments are easy for banks to fund on balance sheet. The bank can swap its short-term deposits for medium maturity fixed-rate liabilities. Or it can use corporate or covered bond markets to issue medium-term fixed rate debt.

However, the author also cautioned that this funding approach can have implications for mortgage design as well –

> outside the United States almost all corporate debt is non-callable. Thus, a lender using covered bond or noncallable corporate debt will incorporate a pre-payment penalty in order to maintain a relative match with its funding. The importance of pre-payment penalties has increased with the strengthening of asset-liability matching requirements in European covered bond legislation. Nearly all such legislation requires strict matching with requirements to match balances, coupons and cash flows between the cover pool and bonds.

Moreover, covered bond legislation may also restrict the loan to value (LTV) ratios.

Table 8.4 shows the growth rate of loans for house purchases across major European countries. Over the period 1999–2007, Spain, Ireland and Italy recorded almost 20% and above growth rates. It is important to note that several of these high-growth countries ran into subsequent major crises.

In the UK, there have been significant growth in ARMs. According to Miles (2004), the relative cost of long-term finance may have been a significant factor in the UK consumers'

Table 8.4 Growth rate of loans for house purchase

	2000	2003	2005	2007	Average 1999–2007
Belgium	10.6	14.5	17	10.5	11.5
Germany	4	1.7	1.3	-0.9	3
Ireland	24.1	24.5	28.1	13.5	23.4
Greece	27.1	26.3	36.4	25.2	30.3
Spain	21.5	21.6	24.3	13.1	19.8
France	6.6	9.6	14.8	12.7	10.1
Italy	20.7	18.2	17.7	12.2	20.3
Luxembourg	22.7	17.6	13.4	22.9	14.1
Netherlands	24.6	12.3	13.5	2.5	13.4
Austria	11.1	10.4	12	6.9	13.2
Portugal	20.4	8.1	15.8	9.2	14.9
Finland	10.6	16.4	16.7	12.4	14

Source: Based on data from ECB; for a full list and discussion, refer to the source. https://www.ecb.europa.eu/pub/pdf/other/housingfinanceeuroarea0309en.pdf

preference for ARMs. The Miles report (2004: pp. 1–2) pointed out several aspects of mortgage financing[3]:

- "When choosing between mortgages a great many households attach enormous weight to the level of initial monthly repayments. Consideration of where short-term interest rates might move in the future, and of what this implies for affordability, seems to play a far smaller role than it would if households considered the likely overall costs of borrowing over the life of a loan. Average loan to income ratios on new borrowing have risen greatly in recent years; the dangers of more and more borrowers taking on debt that may be manageable at current interest rates, but where affordability could become a real problem should interest rates move up by even relatively modest amounts, are real."
- The report also notes that the risk characteristics of mortgages are subtle and complex. Many households may find it challenging to evaluate these risks to their benefit and with a lack of sound advices, the problem gets exacerbated.
- The structure of mortgage pricing can generate cross-subsidisation from many existing borrowers to new borrowers. Many existing borrowers pay standard variable rates (SVR) while most new borrowers opt for discounted variable and short-term fixed-rate mortgages. This creates unfairness with deteriorating market transparency. "It plays to a tendency of many borrowers to focus on the initial monthly payments on mortgage and it makes medium-term and longer-term fixed-rates appear expensive."
- "Many lenders feel that they are severely constrained in the type of charges they can make for early repayment of fixed-rate mortgages; understanding of the nature of some types of charge amongst borrowers, and of the rationale for any charges, has not been high and as a result types of mortgage that might be suitable for many households are not offered."

There are significant institutional differences in the mortgage market across leading economies in terms of mortgage equity withdrawal, refinancing option, typical LTV, average tenure length, per cent of covered bond and MBS issues. A number of countries allow mortgage equity withdrawal. However, a large number of countries do not encourage refinancing. Typical LTV is about 75–80% with 25–30 years tenure. As noted before, Denmark stands out in terms of covered bond issues and the US stands out in terms of MBS issuance. Table 8.5 shows the variable rate loan characteristics across a selection of countries.

A combination of government support, internationalisation of the mortgage funding, ability of international investors to invest across the global locations and the advent of complex, securitised financial products and services have led to significant ease in households' access to mortgage financing, which in turn have led to households having significant debt obligations. The growth in households' debt obligations have been phenomenal across most major economies. Within a span of 10–15 years, the numbers have gone up significantly. One IMF report notes that the median household debt-to-GDP ratio among emerging market economies increased from 15% in 2008 to 21% in 2016, and among advanced economies it went up from 52% in 2008 to 63% in 2016. A very high portion of the countries report more than 70% of their GDP as household debt. In the emerging economies, while still reporting significant

---

3 Miles, D. (2004) The U.K. mortgage market: Taking a longer term view. Final Report and Recommendations, U.K. Treasury. http://webarchive.nationalarchives.gov.uk/20050301211422/http://www.hm-treasury.gov.uk/media/BF8/30/miles04_470%5B1%5D.pdf

Table 8.5 Variable-rate loan characteristics

Country	Type	Caps	Margin	Period	Options	Discount
Germany	Reviewable	Rate of insurance policy available	N/A	Lender discretion	Mixed	
UK	Reviewable; indexed (tracker)	Caps and collars available (tracker)	0.5–1.5% to base rate	monthly		up to 1%
Canada	Indexed; prime rate	Indexed; prime rate	–0.5%	With prime change	mixed; conversion	yes
Australia	Reviewable	none	1.2–2.2% average spread-to-cash rate	Lender discretion		–1%
US	Indexed; hybrid	Yes; periodic, life of loan	2.50%	1 year; 3:1, 5:1	Conversion	yes
Japan	Indexed; prime rate	Payment cap associated with flexible term		6 months	flexible term; conversion; mixed	on rollover 1–2%

Source: Based on data from Lea, M. (2010) International Comparison of Mortgage Product Offerings. Research Institute for Housing America. Note some aspects may have changed in recent years but this gives a view of the variability across some countries. For a full list and discussion, refer to the source. https://cbaweb.sdsu.edu/assets/files/research/Lea/10122_Research_RIHA_Lea_Report.pdf

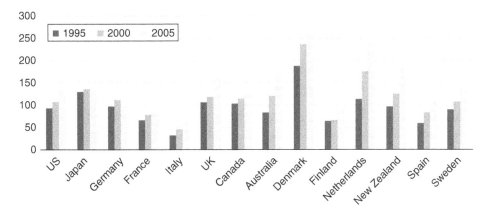

Figure 8.2 Household debt per cent of annual disposable income.
Source: Based on data from OECD. http://www.oecd.org/eco/outlook/39698857.pdf

growth, trend is more tapered. Expansionary monetary policy may have attributed to these changes. But, high household debt obligation can have the financial stability risks.

Such high growth has also led to households having a high portion of their annual income as the debt obligation. Figure 8.2 report household debt as a percentage of annual disposable income for three representative years: 1995, 2000 and 2005. Within a short span of about a decade, many countries have seen this metric rising above 100%. While this can bode well for aggregate expenditure and growth in homeownership, this is alarming. The underlying weakness of the countries can get amplified in the face of financial turmoil, as experienced in 2008. Within Europe, Ireland, Spain, Mediterranean and Central European Economies (CEE) had the biggest booms and crashes. The Eurozone crisis in recent years has also amplified the housing market downswings in some countries, while the Scandinavian countries have done relatively well. Europe shows a very diverse housing market with much difference in the social and economic system, which are also reflected in the mortgage financing market. While being in the same trading bloc might encourage integration and the ability of frictionless investment flow can enable such integration, there is still much divergence with significantly varying regional patterns. For example, if we take a look at two major European economies – UK and Germany – we see a marked difference. UK house prices have been more volatile than German house prices. There are some fundamental differences between these two countries such as in homeownership rate as discussed in the previous chapter, the social housing provision. The UK housing supply is a lot more supply-constrained than that of Germany. With decades of development in planning the regulation and governing system, the UK has a very inelastic supply compared to Germany, which is one of the big reasons behind long-run price rises. With long-run price rises, the housing financing becomes even more crucial to fund house purchases. There are also significant differences in the social system and attitude to savings. As discussed in the previous chapters, marginal propensity to consume and save (MPC and MPS) are quite important aspects of housing demand and demand for housing finance. German consumers tend to have a higher preference for savings and an aversion to borrowing. Even the credit card usage is lower compared with the UK. Such attitude also leads to attitude to homeownership. The development of leverage-led housing is limited in Germany and it is common to use long-term fixed rates. Arbitrage possibility, the rampant use of a refinancing option and frequent and heavy withdrawal of housing equity to finance other high-value consumptions are limited. In general, the mortgage finance system crucially depends on the attributes of the underlying housing market and consumer attitude towards loan financing.

Figure 8.2 presents household debt as a per cent of disposable income over three time periods. The rise over a decade in a household's reliance on leverage is remarkable across all countries.

In this chapter, I have discussed the housing finance market and associated developments. With the global financial system and international investment platforms, housing has become an important asset class in terms of investment potential. Investors can not only participate in the secondary mortgage market but also in more direct housing market channels. Moreover, digital technology and other hard technology innovations are opening up new areas of investment. Therefore, I discuss the issues for the residential real estate as an asset class in the next chapter.

Selected research topics

- Factors explaining country variation in mortgage market depth.
- Role of mortgage risk in business cycle fluctuations.
- Mortgage market and housing affordability.
- Role of technology in mortgage market.

9 Residential real estate as an asset class

Chapter outline
- Comparison of residential real estate as an asset class
- Student housing
- US, UK, Australia
- Market for investment in residential real estate
- Cross-border investment in local housing markets
- Emerging trends – Fintech, Proptech, AI in RE/housing
- Selected research topics

Housing is an important asset class from all perspectives. For a household's perspective, it is perhaps the most significant vehicle of creating wealth. From a retail investor's perspective, housing in many countries has given strong long-run returns. Housing can earn significant capital return as well as income return, especially in countries' rising house prices and rent. Due to supply constraints across most parts of the market in several countries (including the advanced economies and specific locations in emerging economies), prices (house value and rent) have been rising. In this chapter, I highlight opportunities, issues and challenges of housing as an investment asset class, drawing from the relevant literature.

There are many sub-sectors of the housing market that are of significance as an asset class: investment through mortgage market, investment into the rented sector and niche markets (student housing, retirement housing). Moreover, foreign or cross-border investment into the housing market across most well-functioning housing markets has been strong over the last couple of decades. At the same time, with the advent of digital technology, the application of robotics, artificial intelligence (AI), machine learning, block-chain technology along with a huge amount of data generation, there are now significant opportunities to invest in residential real estate using Proptech platforms.

There are significant challenges that can make the residential sector less attractive to the investors. Regulations, government interference (political economy aspect) in the form of taxes and subsidies, supply-side constraints and dependencies are of huge significance as an asset class and create frictions that the investor community does not typically enjoy. However, if the investor is able to manoeuvre these and have a support system to deal with issues, the sector can offer a sizeable opportunity, as housing is a need for all citizens. Complexity of the issues themselves can sometimes open up opportunities for investment.

Investment through the mortgage market has been dealt with in detail in Chapter 8, referring to channels of secondary or securitised markets. Investments may involve both equity and debt-financing. A debt investor can earn an interest income by investing directly in the asset or can invest through financial vehicles such as mortgage-backed securities. Debt investors may not receive any ownership stake like the equity investors, who can have an ownership stake. Generally, equity investments offer higher return due to the presence of risk premiums associated with the ownership. So, there are several ways investments come to the housing market: direct investment into the asset, through financial vehicles that are purposed for investing into the housing sector such as funds managed by professional fund managers in the indirect market, buying stocks of companies who invest in the housing market or homebuilders, equity release and sharing, investment in residential land, house price derivatives, etc. There are more and more calls for involvement of private sectors for housing supply solutions and funding.

Overall, the big aspects of the housing market that can create significant investment opportunities are:

- Lack of supply: There is a severe lack of supply in many parts of the world. Supply of suitable housing is very constrained. Regulations as well as funding issues related to large-scale developments are significant barriers for the housing supply. Various public sector agencies and private sector organisations are involved.
- Urbanisation: as mentioned before, with the world's urban population expected to rise to almost 70% of the total population by 2050, urban areas throughout the world need a significant amount of housing.
- Population growth and changes in demography: population growth within certain demographics are fuelling demand for bespoke housing such as the ageing population and the need for retirement housing.
- Rising income level: Income levels across some developing regions are rising due to high growth rates in countries such as India and China. A large section of the population experiencing rising income levels can fuel aspirations of homeownership, creating a huge demand for housing.
- Rising housing costs or worsening affordability: several parts of the world, especially in developed economies and some urban areas, are experiencing fast-rising housing costs leading to a severe dent in affordability.
- Needs for innovation: while there are needs for traditional solutions to housing issues, the market is also opening up for innovative solutions including new designs for usage of space, use of new materials in housing construction, new models of housing provision and use of new and digital technology (e.g. 3D printed housing units or sections).

Private renter sector

Due to rising costs of homeownership, changes in lifestyle and the attractiveness of town centre living, there is a high demand for private rented sector (PRS) units. The flexibility of housing occupation as a renter has always attracted some people. Coupled with other drivers, the PRS market is ripe for investment and innovation. Regulations, especially in the 1990s, related to rent control and also on the tenant services, along with the rising popularity of homeownership, had shifted the focus away from the PRS for the large, institutional investors in several countries. A prime example is the UK. After several decades of declining PRS share, the market started to be more active in the recent decade. Unaffordability and lack of supply in the urban areas have further fuelled the need for creating an environment for large-scale investors to return to PRS investing. Moreover, the financial market has also changed, with more structures and vehicles available for such investment. PRS does offer both capital appreciation and good income return, especially in certain locations. While buy-to-let activities have fuelled some of the recent growth in PRS, the sector really needs professional investment that can reap the benefits of economies of scale and offer services to tenants that are scalable, away from the small-scale, amateur landlords. Purpose-built, large-scale development can close on a consumer need and offer aspects of PRS that small-scale landlords cannot. Longer-term tenancies, more predictable rents and shared services and well-functioning communal shared spaces (such as a concierge or rooftop garden or in-house cafeteria) and more importantly professional services and management of daily maintenance issues are more likely to be delivered by large-scale operation.

First-time buyers are facing high barriers to becoming homeowners or getting on the housing ladder due to much divergence between average earnings and house prices along with the issue of significant levels of deposit constraints. Moreover, mortgages that are typically on offer to the first-time buyers are not cheap. Such unfavourable market dynamics for young, first-time buyers are boosting the demand for PRS. However, there is severe lack of supply of quality units in rented accommodations. Quality is often poor, with damp conditions (leading to poor indoor air and environment quality), low quality furnishing and fixtures, small-scale landlords failing to address issues quickly. The poor quality also comes with insecurity in tenure due to short-term tenancies with annual rise in rent. Many tenants frequently cite such issues as their reasons for leaving.

Ball (2007), in research sponsored by the Investment Property Forum, notes several ways that large investors may be able to derive gains from residential property[1]:

- Good returns: Residential investments can offer good total returns. With rising house prices, the capital values can be preserved and appreciated along with income return.
- Weakly correlated returns: Portfolio construction and diversification can be easier in the residential market as returns tend to be imperfectly correlated and weakly correlated across various spatial scales.
- Specialist skills add value, where they matter: due to changes in consumer preferences, specialisms can be maintained as offering unique solutions to consumer needs. Large-scale investments can effectively look to exploit such benefits.
- Average rental returns are likely to be attractive: There is less uncertainty and risk from investment cycles in housing when the supply is severely restricted, which would require long-term fixes.

Stable, long-term returns can attract pension funds and other large investors. They can either buy up blocks of existing rental accommodation or can undertake development options such as 'build to rent' (similar to those in the US, Canada and continental Europe). While buying up existing accommodation is attractive as it does not require going through planning and other developmental processes with much uncertainties and delays, it can limit the offering. New development allows the investors to offer new, innovative and attractive solutions to the market, which can significantly raise the likelihood of sustainable revenue and profit. Moreover, with fast-changing regulatory environments in terms of energy efficiency and digitalisation, the new developments can help avoid obsolescence and costly refurbishing needs.

Many governments frequently take up schemes and measures to bolster investment in the housing market. PRS is an area which requires such moves. The regulatory environment needs to be investor-friendly. Ease of investment, liquidity, favourable tax regime, safety of

---

1 Ball, M. (2007) Large-scale investor opportunities in residential property: An overview. Investment Property Forum, UK. http://www.ipf.org.uk/resourceLibrary/large-scale-investor-opportunities-in-residential-property—an-overview—november-2007-.html

investment, certainty in the political climate, lower information asymmetry etc are important from the investment perspectives. In the UK, the government has unveiled a series of measures aimed at boosting institutional build-to-rent, such as a £1bn equity fund to trigger development, £10bn of loan guarantees for the social and private rented sector, new draft planning guidance on build-to-rent projects and a governmental task force overseeing the fund and loan guarantees. Such measures do send positive signals to the investment community. Moreover, these can also be viewed positively by the tenants.

However, there are significant problems associated with direct investment in the Private Rented Sector (PRS) (Ball, 2007). It is important to note that some of these problems are inherent to the housing market and little can be done in terms of alleviating those.

- High transaction costs: It is likely that the costs of direct investment are relatively high in the housing market. Combining different fees and non-monetary costs, it can be substantial and act as a deterrence for investment. Any potential for low return due to high transaction costs can dissuade many investors getting into the housing market. The availability of alternative investment channels which offer lower cost is plenty, for example, stock, bonds and commercial real estate.
- Significant management and maintenance costs: residential tenants tend to be highly demanding in terms of maintenance. Unlike commercial properties where tenants can be responsible for those issues, housing tenants can and do demand regular maintenance and upkeep of the property. Such demands can be costly, labour-intensive and tricky. Even if the services can be dealt with by third parties or outsourced, there is still significant need of managing those and resulting in high costs. The service can be labour-intensive, which not only adds to the costs but also create challenges of managing a professional in-house or on-demand labour force. At the same time, any delay in providing services or low quality services can get easily escalated by the residents.
- Turnover and vacancy: due to short-term tenancies, popularity of homeownership and contracting conditions, the turnover rate is very high in the PRS. Breaking of a tenancy is not difficult. Tenant mobility tends to be high. Re-tenancy is often costly due to maintenance work and search costs. The high mobility can also lead to high rates of vacancy, which not only undermines income return but also have implications for maintenance of vacant properties. Countries where long leases are allowed and popular, such issues tend to be of less significance such as in Germany.
- Reputation: issues of laxity in services, delay in returning deposits, unfortunate incidents etc can make a huge deal in the PRS stock. The reputation risk is high and can make investors nervous about investing into PRS. However, large-scale operations can be susceptible from the damage from such issues. Especially with the existence of other alternatives, this can drive out the tenancy demand from one landlord to other. Managing such reputation risks is tricky as well.
- Illiquid, thin markets: although compared to commercial markets, residential real estate experience a lot more transactions, the number of comparable sales is still low, to be able to benchmark and re-price appropriately. Difficulty in pricing is quite problematic for investors looking for capital value appreciation.

Despite the problems and challenges involved, PRS can be an attractive market for private, professional and institutional investment. The opportunity is significant across all markets. However, investing into PRS may require specialist skills and knowledge of the sector for long-term, sustainable gains.

Student housing

Global investment into student housing reached $16.4bn in 2016, setting another annual record, according to a report from Savills[2]. Huge amounts of investment have been reported in several major education markets such as in the US, UK, Germany, France and Australia. The US experienced almost $9.8bn total investment in 2016 in student housing. According to the Savills report, annual investment in the sector increased 245% in France and some 380% in Germany in 2016/17. Global cross border investment into student housing was also quite high, standing at 37% of all investments. This is a higher sector proportion than for offices (34%) and retail (29%). Many sovereign wealth and pension funds are attracted to the sector.

What drives student housing?

The number of students entering college and university or vocational education is very high across all countries. This is fundamental, as the role of education and the requirement of a formal skillset and degree labels are important for employment prospects. Employers frequently look for a formally trained labour force with recognised providers. Therefore, there is much more emphasis on college or university education. The trend is global. Some of the large countries such as India and China are producing huge numbers of graduates. The graduates are also able to find suitable jobs due to availability of specialised jobs and service sector innovation. For example, the information technology sector in India requires a huge amount of skilled workforce due to the shift in the industry and rising domestic as well as global demand for IT-related products and services. This has fuelled proliferation of the providers as well. With high demand and also a better supply of quality institutions, the student number has gone up. Many students often have to move out of their home towns due to location of the education providers being concentrated in large metropolitan areas and employment centres. This means it creates a huge demand for student housing. Availability of in-campus accommodation and private accommodation with easy access to the campus are often important considerations for the choice of providers.

Moreover, over the last several decades, due to the availability of globally-recognised providers and ease of international travel along with favourable immigration policies, the number of students moving internationally has gone up significantly. In any global university ranking such as QS, Times Higher Education and CWUR, the share of US, UK and other countries is very large and those are the most favoured destinations of international students. Western education, particularly in the US, UK, Canada and Australia, is favoured by the students as well as employers. Moreover, these English-speaking countries and institutions offering education in English language is an important factor for the majority favouring education in English as a global language.

At the same time, active internationalisation strategies for higher education have been undertaken by global institutions to attract talents and boost revenue. Per capita income growth, availability of the information network, government support and family support in large emerging economies have fuelled much growth in international student numbers. Each year, hundreds of thousands of students move to leading education destinations such as the US, UK, Australia, Canada and Continental Europe from emerging economies like India, China and African countries. For example, according to Savills research, China, which is the largest outbound market, accounts for as much as 17% of all international students globally. These numbers can grow as the outbound mobility rate is still lower than those from other countries. China's outbound mobility rate is 18.5 students in every 1,000, compared to the

---

2 https://pdf.euro.savills.co.uk/global-research/world-student-housing-2017-18.pdf

mature markets of France (33 per 1,000) and Germany (39 per 1,000). This will create a huge demand for student housing.

Supply of student housing

The availability of quality student housing is low and at the same time, tastes and preferences for students are changing with new requirements (such as connectivity, social network, communal space). Students typically look for private space, shared space, easy access to key amenities, good and functional study space, 24-hour and everywhere digital connectivity and locations close to places of interests.

Purpose Built Student Accommodation (PBSA) is becoming common. It is housing specifically built for students and often contracted by the education providers and built by commercial developers. This does not include the private rented sector focused on student accommodation. PBSA offer several aspects of student-preferred attributes such as high-quality accommodation, rents inclusive of all utility bills, varying tenancy lengths and professional management and security along with locations that are close to the education providers and city centre and other amenities. Ensuring all of these is not easy, but that is the order of the day.

Figure 9.1 shows tenancy and room type by source market. Figure 9.2 shows the availability of purpose built student housing (PBSA) across major markets. The number of beds as a share of total enrolled students is low across the countries.

What drives investment into student housing?

Within the niche or alternative or non-mainstream property sub-sectors, student housing is a significant one. In fact, it can be argued as mainstream due to its yield offerings and attractiveness as an investment. UK and US markets are very attractive from the yield perspective. Several reasons can be put forward behind student accommodation as a viable and profitable asset class within the property sector.

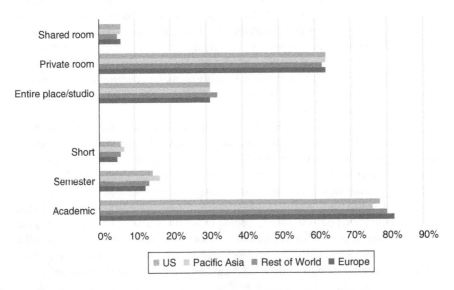

Figure 9.1 Tenancy and room type by source market; 2016 or latest available.

Source: Based on data from Savills Global Research, Student.com; for a full description and list, refer to the *source*. https://pdf.euro.savills.co.uk/global-research/world-student-housing-2017-18.pdf

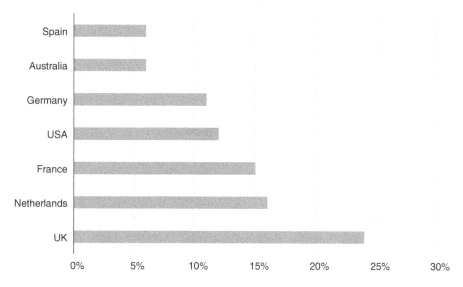

Figure 9.2 Availability of purpose built student housing (PBSA). Beds/total enrolled students (full time and part time); 2016 or latest available.

Source: Based on data from Savills Global Research, Student.com; for a full list, refer to the *source*. https://pdf.euro.savills.co.uk/global-research/world-student-housing-2017-18.pdf

- Student housing can ensure predicable income return. Due to the nature of annual turnover and quite accurate forecasts on the student number, it is possible to be certain of the demand for student housing. This enables preparedness and supply readiness. The certainty in income provides ease of structuring the investment terms and conditions.
- Student housing provides a significant shield from economic fluctuation due to steady demand for education. Changes in the immigration policy can somewhat impact such steadiness in demand from international students.
- Student housing can be standardised in terms of offering. It does not require too many variations in products and services.

Figure 9.3 shows the yield from student housing compared with other asset classes across major locations.

There are global operators such as Global Student Accommodation Group (GSA), Greystar Student Living, Campus Living Villages, the Student Hotel, etc. These global operators have developed appropriate business models for this sector.

Several ways of delivering student accommodation have emerged[3]:

- **Management only providers** – within this model, student accommodation providers partner with the education providers such as universities to provide marketing and operational management services at new or existing facilities.

---

3 See a report on this. http://www.nortonrosefulbright.com/knowledge/publications/130703/the-growth-of-student-accommodation-as-an-asset-class-in-australia

186 *Residential real estate as an asset class*

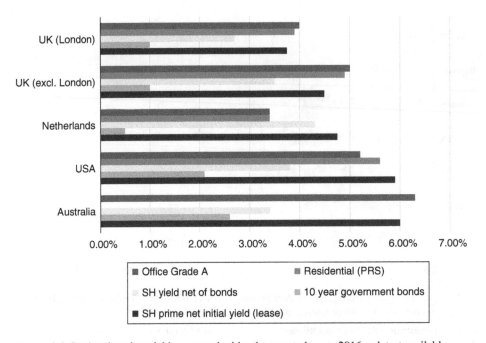

Figure 9.3 Student housing yield compared with other asset classes; 2016 or latest available.

Source: Based on data from Savills Global Research; for a full description and list, refer to the *source*. https://pdf.euro.savills.co.uk/global-research/world-student-housing-2017-18.pdf

- **Build, own, operate transfer** – student accommodation providers partnering with the education providers with a long term lease to build, own and operate the student accommodation and at a contracted time period, the facility is transferred to the education provider.
- **Develop, strata and management** – The units are sold to individual investors on the basis that they may only be used for student accommodation and are leased to the student accommodation operator.
- **Wholly integrated providers** – providers develop, own and manage the student accommodation themselves. Such providers generally buy developable land near to the education provider and deliver the housing entirely on their own.

Retirement housing

I have discussed the scope and issues of retirement housing in Chapter 8 – please refer to the relevant sections.

Cross-border investment in housing

All sub-sectors of housing markets, especially in key global locations, are increasingly attracting cross-border investment. Besides financial vehicles and mortgage-backed securities, direct investment from foreign sources has poured into the ownership of individual units, PRS, student housing, retirement housing, etc. According to a report from the National Association of Realtors, US home sales to foreigners surged 49% between April 2016 and March 2017, to a record $153 billion[4]. The United States is a prime target for foreign investment. Foreign investors from China ($31.7 billion), Canada ($19 billion), the United Kingdom (9.5 billion), Mexico ($9.3 billion) and India (£7.8 billion) poured into the US housing market location between April 2016 and March 2017. Similarly, a large proportion of residential sales in London has been completed by foreign buyers. A large number of new homes in Westminster, Kensington and Chelsea and the City of London are owned by overseas investors. Not only London, other key UK locations have experienced such an influx of foreign investment. The UK property sector tends to be particularly popular with investors from Asia such as Hong Kong, Singapore, Malaysia and China.

These are staggering numbers. The question is: what drives such volume of investment?

- Retail buyers purchase properties in foreign locations for many reasons – live, work, invest or permanently relocate. Often, family connection plays an important role.
- Investment tends to be location-specific. Not only preferred cities but also sought-after areas of the city are the main targets. For example, London, Manchester in the UK, major cities in California, Florida and Texas in the US.
- Some purchases are motivated by foreign exchange benefits – currency arbitrage is a leading reason.
- Historical relationship between countries such as past colonial relationship and geographic as well as economic proximity are significant factors for foreign investment.
- Moreover, trophy properties (such as apartments in prominent blocks) are of interest to the foreign buyers.

Impact of foreign investment in the housing market:

- An immediate impact is the restricted supply of properties available to domestic buyers. If a large portion of the stock is acquired by foreign buyers, then a significant supply constraint can emerge. In already supply-constrained markets such as in the UK, this can have significant impacts and potentially exacerbate the issues.
- Foreign buyers may bid up the property prices. This can lead to rising property prices and worsening affordability.
- Some locations experiencing rising and high property prices along with significant interest and activity of foreign buyers may see gentrification and displacement of local people. Major tourist locations and prominent areas are examples of such scenarios.
- At the same time, foreign buyers can provide much-needed funding for new build schemes with much benefit to the homebuilding sector and to the national and local economy by fuelling job growth and income growth along with significant fiscal revenues for local government.

However, across several global locations, there has been much debate on restricting foreign investment in the housing sector. The housing sector is much more politically active than other parts of the property market due to housing's importance to the society and its effect on all citizens. Low or middle-income households being crowded out of the market due to foreign buyers can be a powerful political argument against such investments. Political arguments often lead to imposing of additional taxes or an altogether ban on foreign buyers

---

4 https://wtop.com/business-finance/2017/07/foreign-investment-in-us-housing-booms/

depending on the political regime that is in power. For example, there has been discussion on imposing additional tax on foreign buyers in London. The incoming New Zealand legislative changes emphasise the banning of foreign ownership of residential property. Australia's Foreign Investment Review Board (FIRB) already applies a ban on foreign ownership for existing (as opposed to new) housing. The economic argument is that removing or restricting investor demand by banning foreign nationals will help stabilise house price growth. Switzerland has imposed a quota on the number of residential properties that can be sold to foreigners. Vancouver in Canada has recently introduced an additional property tax on foreign home buyers. Much of these policy moves can be driven by a political agenda – nationalist regimes tend to have much stronger views on these compared to liberals. Rising nationalistic politics can fuel the imposition of various types of measures to restrict foreign investment. This can change from time to time, depending on the economic and political climate.

Sá (2016) analysed the effect of foreign investors on local housing markets. Using data from the UK, she reports that the share of total transactions registered to overseas companies (in volume and in value) show a similar upward trend to house prices over 1999–2014 and there are considerable regional variations in foreign investments with a concentration mostly in the South East and major cities in the North, such as Liverpool, Leeds and Manchester. The study used regional variation in house price growth and the share of residential property transactions registered to foreign companies to estimate the effect of foreign investment on house prices. The study also applied an instrumental variable to address endogeneity concerns with respect to location choice. There are several interesting findings:

- The main finding is that an increase of one percentage point in the volume share of residential transactions registered to overseas companies leads to an increase of about 2.1% in house prices.
- The study also reports a 'trickle down' effect to less expensive properties.
- Using house price-earnings elasticity for each local authority in England, it is found that foreign investment increases prices more in local authorities with a larger house price-earnings elasticity compared to those with a less elastic housing supply.
- The study does not find any statistically significant evidence that an increase in foreign investment may lead to an increase in housing construction. One of the arguments often quoted in popular media that foreign buyers purchase properties purely for capital appreciation and keep them vacant is also investigated but did not seem to have much statistical significance.
- The study finds, though, evidence of detrimental effect on homeownership rates, suggesting that some residents may have been displaced or priced out of the market in areas where foreign investors are more active.

The impact of foreign investment can be mixed, having some adverse and some favourable effects. It depends on the contexts within local areas to determine such impact.

Opportunities for investing in technology for residential real estate

Since the widespread use of the internet, the property sector has been undergoing slow but steady changes with each and every wave of digital technologies and associated potential for innovation. Real estate is slow to adjust to changes due to the complex nature of the asset class and involvement of several actors and organisations in the market. The recent wave of digital technology is significant and it has potential to make fundamental changes to how we do things. Real estate – both commercial and residential – is very prone to this wave. This is now widely called PropTech. PropTech can encompass a broad area of technology use – right from the building technology (smart buildings, IoT-based (internet of things), smart city development (more in Chapter 10), financial aspects, 3D technology and augmented reality to data analytics.

Several aspects of the technology can potentially lead to effective, far-reaching innovations in the real estate industry. Property management and performance and benchmarking can all be enhanced by technology. The benefits of technology is immense: it can reduce costs; it can increase efficiency; it can reduce delays in the planning processes; it can identify more accurately the needs and areas of development; it can help with accurate performance measurement; it can lower the possibility of wild speculation; it can reach more citizens, etc. Investment into technology that helps with the development of tools, products and services can be very useful and sustainable in terms of profits and revenue.

There are already many platforms and start-ups focusing on technological innovation. Right from 3D printing building sections to better material to digital connectivity, there is opportunity for investment. Real estate development is no longer just in the realm of developers but also technology companies. The borders and realms of 'physicality' and 'digitality' are getting blurred and property market is right in the middle of this change. Many new buildings and developments are now in collaboration with traditional real estate development companies and technology partners.

Technology is also making positive inroads to the transaction process and financing. Applications of block-chain technology and machine-learning have the potential of removing the inefficiencies in the transaction process, which have become almost inherent to the real estate industry. While service propositions are linked with those inefficiencies, the benefits can far outweigh the costs, and it can actually lead to new job creation.

Having said that, there are and will be resistance from some corners of the industry, which should dissipate as alternative job prospects as well as skillsets emerge and get developed. While it is certainly possible that this can and would replace some jobs, it is also beginning to be clear that benefits from technology use can free up human-hours which can be used for productive purposes and thus the long-run benefits would outweigh the short-run cost. Albeit, it is not entirely clear how and what alternative job prospects will be there and until then, the resistance to the technology adoption will persevere and justifiably so. I firmly believe that technology should be used as 'enablers' and should not be a 'replacement', as that would ensure loss of livelihoods and social cohesion. If the alternative skillsets and employment cannot be guaranteed or provided to the section of the society affected by technology, then we do need to invest thoughts and resources in ensuring that. A well-functioning housing market is very important for the society. There is a growing acknowledgement of this view in the industry. Improvement and transformation are important for industry. Viable business models need to be developed with much emphasis on sustainability. One solution can not fit all as the property sector reflects huge amounts of heterogeneity and need a clear and well-grounded strategy to face this digitalisation wave. A few aspects need tightening: the ability to respond and change, a belief in symbiotic business existence and a view towards long-run sustainability.

Another potential area of investment is skill-development and education for the real estate industry. With new technology and big data application, new skillsets will be required for future real estate professionals. There will be a much greater need for basic data science skills and technology skills to be able to interact with and advise the tech-savvy, informed customer base. Investment into those can have far-reaching implications and can indeed enhance the existing returns from the current companies. Hood and Nanda (2018) identified a set of new/emerging skills – design thinking, future thinking, measuring success, balancing art and science, organisational design and understanding value versus cost. All these would require new educational approaches and appropriate tools.

A significant room for real estate investment is opening up for data centres. As a sub-asset class within the property sector, a data centre offers a very sizeable opportunity for investment. Data centres used to be housed, owned and operated by a small number of companies with their in-house needs. A vast majority of companies now need data centres to store huge amounts of data that are being generated by their business operations and customers' activities. With the General Data Protection Regulation (GDPR)[5] in Europe and comparable regulations across most economies, protection and maintenance of data is being taken seriously. This creates very significant regulatory and business risks for the firms. This requires specialised facilities for the data storage and management. There is now a huge demand for data centres. Management of data centres can benefit from some of the real estate principles. Those can be viewed as properties. Less heterogeneity and ability to provide standardised solutions are big benefits. However, with a rise in the number of data centres and their appetite for energy use opens up a whole range of questions, which can be addressed to a large extent by real estate principles.

In Chapter 2, I discussed several unique attributes of the housing market, as mentioned in the literature (see Meen, 2001; Miles, 1994 and others), such as: (a) Territoriality or Spatiality or Immovability or spatial fixity in the housing market; (b) Variability and Heterogeneity in the housing market; (c) Longevity or Durability of housing units; (d) Informationally imperfect market; (e) Transaction costs in the housing market; (f) Externality and external effects in the housing market; (g) Socio-political influences in the housing market. The question is: how will these be influenced by the recent PropTech wave? The following discussion can shed light on this.

In the following section, I deal with the technology aspects in more detail, with reference to application areas along with a brief explanation on the technology platforms.

Some terminologies need explanation within this context.

Big data

As we are increasingly getting connected through many devices, a massive amount of data is being created. All these data points can help understand the processes and patterns better. However, analysis and processing of the massive data requires carefully built tools and techniques. How big is 'big data'? According to an IBM estimate in 2013, 2.5 quintillion bytes of data is created every day (1 quintillion bytes = 1 Exabyte = 10^{18} bytes = 1000 petabytes = 1 million terabytes = 1billion gigabytes.[6] This data comes from everywhere: sensors used to gather climate information, posts to social media sites, digital pictures and videos, purchase transaction records, cell phone GPS signals, etc. Five attributes of 'big data' (according to, originally, Doug Laney and SAS Corp., along with many other industry commentators) are: Volume, Velocity, Variety, Variability and Complexity. Within real estate, these attributes can be easily found along with much heterogeneity collected from both physical and digital environments. It is important to note that a property is not an isolated being, it is surrounded and connected

---

5 The General Data Protection Regulation (GDPR) 2016/679 is a regulation on data protection and privacy for all individuals within the European Union and the European Economic Area.
6 https://www.ibm.com/blogs/insights-on-business/consumer-products/2-5-quintillion-bytes-of-data-created-every-day-how-does-cpg-retail-manage-it/

to many other 'things'. Standard estimation and statistical techniques need to be carefully considered and various methods need to be combined. Enormous data with deep complexities coupled with unobserved heterogeneity pose a formidable challenge, requiring multiple and parallel frames of analysis. For contextual applicability, social scientists will need to work with computer scientists and mathematicians for modelling and making sense out of the 'big data'.

Artificial intelligence

Artificial Intelligence (AI) is about machines that can function like humans. SAS Institute Inc. describes it as –

> Artificial intelligence (AI) makes it possible for machines to learn from experience, adjust to new inputs and perform human-like tasks. Most AI examples that you hear about today – from chess-playing computers to self-driving cars – rely heavily on deep learning and natural language processing. Using these technologies, computers can be trained to accomplish specific tasks by processing large amounts of data and recognizing patterns in the data.[7]

AI research goes back to the 1950s but our ability to process huge amounts of information and modern-day fast computing technology are enabling major application areas in recent years. Google CEO Sundar Pichai in 2018 said artificial intelligence may have a bigger impact on the world than electricity or fire. Regardless of rhetoric, AI is undoubtedly going to add massive abilities to human civilisation that can potentially change the way we are used to living lives, with the appropriate framework for ensuring proper use. As SAS Institute Inc. notes:

- AI can automate repetitive learning and discovery through data.
- AI can add intelligence to existing products.
- AI can adapt through progressive learning algorithms to let the data do the programming.
- AI can analyse more and deeper data using neural networks that have many hidden layers.
- AI can achieve incredible accuracy though deep neural networks.
- AI can get the most out of data. Data is now a very valuable resource.

Machine learning

"Machine learning is a method of data analysis that automates analytical model building. It is a branch of artificial intelligence based on the idea that systems can learn from data, identify patterns and make decisions with minimal human intervention" (SAS Institute Inc).

Neural network

Robert Hecht-Nielsen, inventor of the first neurocomputers, defines a neural network as: "a computing system made up of a number of simple, highly interconnected processing elements, which process information by their dynamic state response to external inputs." Artificial Neural Networks (ANNs) are algorithms and devices that are based on the neuronal

---

7 https://www.sas.com/en_gb/insights/analytics/what-is-artificial-intelligence.html

structure of the human brain and nervous system and can work in unison to solve a problem, for example, an application in designing transportation infrastructure systems.

Deep learning

Deep learning is about artificial neural networks working through many layers of data and the algorithms starting to be 'trained' to learn from each operation and be able to apply in different contexts using the lessons and same principles.

Natural Language Processing (NLP)

Human language is complex, with multiple meanings depending on the contexts. An AI can be trained to interpret human language through programming and training. This is called Natural Language Processing (NLP), for example, AI-based assistants Alexa and Siri.

Blockchain technology

Blockchain technology is a public register – distributed ledger technology, based on which transactions (e.g. financial transactions) between two parties are stored in a permanent way. The information is secure and can be verified using unique path information on the cryptographic blocks (chain of blocks connected by the timestamp and event information) in which data is saved. This is why it is called blockchain. This inherent ability to verify and certify the ownership and transaction process and save all the information is a very powerful and attractive feature. Based on this principle and blockchain structure, bitcoin cryptocurrency has been invented. While financial transactions is an obvious application area, this can be applied in stock market or in other markets. In real estate, blockchain-based smart contracts can be considered; it can also be used for property purchases, sales, financing, leasing and management purposes.

Augmented Reality (AR) is an application area of enhancing the real-world experience within the real environment with computer-generated images and perceptions that can offer alternative experiences and possibilities. Augmentation of the real world objects is based on AI, machine learning and analysis of 'big data'.

The use of AI now is very widespread. Many features of modern computing, smartphones and tablets, are based on AI ability. For example, automatic product search suggestion, customer care through chatbots, Apple Siri, live-tracking SatNav that can recommend real-time best possible routes all use some form of AI and machine-learning capabilities. Many warehouses have started using autonomous units. There is now huge interests and realistic possibility of driverless autonomous cars. There are discussions of AI-based tools in the healthcare sector that can identify illness and symptoms much quicker and more accurately than humans. The ability to analyse huge amounts of information in a real time basis and be able to learn from the processes is the key feature of all these bursts of technology. With the 5G connectivity coming around in 2019, the application areas can be further extended.

The property sector offers many application areas. In a recent report, Baum (2017) has provided a good descriptive account of the current status in this area by interviewing several industry leaders and professionals.[8] The housing sector is a prime example. With

---

8 Baum, A. (2017) "PropTech 3.0: The future of real estate". Said Business School. University of Oxford.

all of us being active in the housing market in some way, the AI application can affect all citizens.

Technology application areas in the housing market

A **Direct technology application in home search:**

In Chapter 6, I discussed the housing transaction and search process. Figures 6.6 and 6.7 show the seller's and buyer's journey in a typical transaction. Many aspects of the role of an estate agent can be enhanced, partly or fully delivered by the new technology. There are significant investments flowing into making the transaction process for both buying and renting easier and friction-free. Online search engines such as Rightmove and Zoopla are prime examples with online listing by the estate agents. More recently, there are platforms offering no-agent transactions, as led by the seller with a fixed fee with fully online-managed services such as Purplebricks. This does not necessarily mean complete. As discussed in chapter 6, there are two major roles of an estate agent: matching and bargaining. The matching aspect is about bringing a suitable house (and the seller of the house) to the buyer to agree on purchasing. Much of that role can be and is being performed by online search engines and platforms. With augmented reality, the speed of internet as well as mobile technology (3G, 4G and soon 5G) can effectively deliver options and experiences of the search process to all parties. Friction-free interactions between the seller and the buyer along with automated processes for predictions and simulation of what-if scenarios can be delivered by current and emerging technology platforms. Huge amounts of investment are pouring in and expected to pour in over the next 5–10 years. It can be reasonably surmised that over the next decade, much of the housing search will be performed in online platforms with minimal human (estate agents or brokers or licensees) interferences. This can not only reduce the monetary transaction cost of a house (which can be almost 10% in the US) but also it can reduce time and make the process worry-free with much benefits in terms of non-monetary costs of a typical housing transaction.

The market share of online brokers or estate agents is on the rise. Rightmove, Zillow, Zoopla and others are taking market shares fast. Several online platforms have established or are emerging in two of the most populous countries in the world – India and China. Beyond just being an effective search engine, these portals collect a huge amount of information on property market dynamics. As discussed before, property market being inherently heterogeneous and informationally asymmetric, the wealth of 'big data' that property search engines are collecting and processing can be a very significant resource for further revenue generation. New services can be offered by using appropriate data science tools and monetisation of data can open up new channels of significant revenue generation.

Some of the areas where the monetisation of the information can happen are:
- Predictive analytics – prediction of market trend, demand and supply shifters, property prices, consumer sentiment. Potential customers for predictive analytics are numerous in number – from sellers/buyers to lending institutions to policymakers.
- Property development – better understanding of taste and preferences of customers can help develop more relevant properties.
- Property valuation services – more granular information available for property can enhance the quality of the benchmarking process. Sellers can get a much better sense of the expected price and the gap between the buyers' offer and sellers'

asking price can be closed down further, which can only help in clearing the market through quicker agreement and less bargaining. This can immensely help with controlling speculative activities and formation of unfounded expectations. The problems of moral hazard and adverse selection in the housing market can be reduced or controlled better. All parties in the housing transaction eco-system (buyers, sellers, intermediaries, lenders, policymakers etc) can benefit from better, more accurate valuation of properties.
- Policy impact analysis – much more accurate analysis of impact of various housing market policies can be undertaken.
- Development of fee-based services for the associated sectors – from building services to the domestic energy sector to the consumer sector can benefit from the property level information and understanding of preferences of their customers.

B **Spillover from FinTech Innovations:**

Financial technology – FinTech – is a sector which sits at the intersection of financial services and technology. Various applications of digital technologies come to play in the financial services industry. With the global integrated market, huge consumer finance market, the application areas are numerous and the opportunities are abundant. Mobile money, internet banking and international money transfers fuel the development of the use cases for the industry. A PWC report (2016)[9] identifies four types of players in the FinTech ecosystem, the As, Bs, Cs and Ds:
- As are large, well-established financial institutions;
- Bs are big tech companies active in the financial services space;
- Cs are companies that provide infrastructure or technology for facilitating financial services transactions such as card payment services, transfer channels, exchanges;
- Ds are disruptors: fast-moving companies, such as startups, focused on a particular innovative technology or process.

There are function-based FinTech companies such as those specialising in lending services with technology-based credit evaluation and disbursement, payments and billings, wealth management, international and domestic money transfer.

With more and more financialisation of housing market products and services, especially in the primary and secondary mortgage market activities, several aspects of the FinTech development can effectively inform the financial side of further PropTech development. However, as Baum (2017) notes, PropTech should not be seen as a subset of FinTech. PropTech is much broader than FinTech as explained before.

C **New ideas of physical space use, shared property economy**

As discussed before, a couple of attributes of property market make it absolutely unique – spatial fixity and territoriality. Several uses of physical space are common for daily lives – home, workplace and entertainment. However, with technology, much of our space of daily use can not only be enhanced and made 'intelligent', new uses can be put in place with revenue-earning capabilities. There are now several uses that are based on shared economy principles. Shared uses and use of spare space are becoming popular. With appropriate revenue model, these can be made viable commercial offerings. These tech-driven new and shared uses are very much a part of the PropTech. New generations are much more

9 https://www.pwc.com/us/en/financial-services/publications/viewpoints/assets/pwc-fsi-what-is-fintech.pdf

comfortable with these concepts of sharing spaces for additional revenue. However, the key is to use technology to match the tastes and preferences and also for business development.

As I have discussed the housing issues of the current and future young people in Chapter 8, some of these shared use concepts are becoming popular due to compulsion, which is the inability to become a homeowner due to the unaffordable housing market in key urban locations, which are also key employment locations for the young. But this is not new. In the housing market, there is always some spare space. What is new is that there are now technology platforms that can enable sending out the information of those spare, vacant spaces and perform the peer-to-peer matching much more effectively in a cost-efficient manner. Consider the under-consumption of a large portion of the housing market, especially occupied by the elderly population. Those stay under-consumed due to lack of suitable matching. It is not unreasonable to assume that some of these homeowners would not mind renting out those additional vacant spaces to young members. This would not only provide revenue but also a much needed solution for isolation or loneliness. Having said that, there are issues of trust and security. However, technology can be used for addressing much of these concerns and match suitable co-occupants. Co-living can at least address some of the supply constraints, especially in urban areas with a high cost of housing. Several common issues of sharing can be effectively solved by technology – the use of amenities and billing. A prime example is Airbnb. It has become quite popular. Although there are emerging concerns of security, a high rate of physical depreciation and uncertainty of revenues, those can be solved and are being addressed by technology solutions, appropriate background checks and revenue model. While Airbnb brings in more tourism to some neighbourhoods and communities it can have some adverse impacts, such as moving stock from long-term tenancies to short-term rentals, which can be more lucrative. Several cities have started putting restrictions on this tenure arrangement. Displacement of local residents can be substantial, especially in good locations for Airbnb demand. Tenancy turnover can be high. Airbnb can also make rents rise and since housing supply is heavily constrained in certain areas, it can lead to worsening affordability. It can be transformative for some neighbourhoods and communities in both positive and negative ways.

Similarly, a flexible workplace is another area of much interest. With an increasing number of nomadic workforce and younger generations preferring flexible working arrangements and becoming their own 'SMEs' (small and medium-sized enterprises), the need and potential for 'anywhere-workplace' is huge. Similar to Airbnb, a workplace solution can be devised. Spare spaces in homes, or spaces during non-occupying daytime hours (when the householder is away and children are in school) can be used for flexible workspace. Again, connectivity, security, appropriate matching and revenue models are key and can be addressed through technology. Like homes, any spare space (in shopping malls, sports clubs, etc.) can be used. However, in all these concepts it is important to note that the underlying factors of changes in workplace requirements and attitudes towards are quite dynamic. Hood and Nanda (2018) pointed out several challenges and opportunities for the organisations. A more complete understanding of the issues will require a transdisciplinary approach with multi-disciplinary inputs, as the authors have noted: "It is no longer just about working; it is about living."[10] Technology can facilitate the transformation.

The shared space concept has now raised huge investment interests. Companies such as Airbnb, WeWork, LiquidSpace and several others have made successful ventures into this concept. What lies in the future is not certain. What is certain is that the economics of spare space is sound and can support the shared economy sustainably.

D **Smart contracting of property sales**

As discussed before, the blockchain framework is a powerful way of recording and managing transactions. Being decentralised and 'trustless', distributed ledger technology based on the blockchain structure can effectively take out some traditional centralised sources, which are expensive in terms of monetary and non-monetary costs, and has the potential for abuse and misuse. While blockchain has now become popular for financial assets, can it be applied in residential real estate? In principle, there are aspects of a transaction process which can benefit from blockchain application. In practice, though, there is resistance due to the heavy presence of intermediaries. Regardless, forces of change are at work in convincing those resisting parties. Several areas can be identified: direct buyer-seller contracting, land acquisitions, registration and titling, all environmental and title searches. In Chapter 6, I have discussed the eco-system of players in a housing transaction. The most time-consuming and frustrating aspect is communication among various stakeholders. Parties hold different information. And there is conflict of interest, which is exacerbated by the information asymmetry. As a result, there is a very large number of cases where the buyer's offer price is accepted by the seller but does not get completed in the process. Similarly, in the mortgage approval process, much information is held in different places and becomes heavily process-driven and creates lots of dependencies. Blockchain can accomplish the residential real estate contracting process. However, there are various concerns, as mentioned by Tapscott and Tapscott (2016): blockchain requires huge amount of power; it can be abused by few powerful parties; it can be hacked, etc. Still, there is no reason why a technology platform, based on clean energy along with a tight governance structure backed by government regulation, cannot bring all these disparate datasets together and make the exchange of information friction-free in a secure, trusted, energy-efficient manner. There is far too much dependency and uncertainty in a residential real estate transaction, especially in onward chain situations.

E **Technology in housing finance market**

Another area of technology application is housing finance. The following aspects of housing finance can be addressed through technology:

- Matching of investors to investment (refer to Chapter 8). In the secondary market, the matching takes time and is fraught with lots of uncertainty driven by risk evaluation and market projections. Speculation is rife and can become costly. Matching and engagement tools shared by fund managers and investors can substantially lower the frictions.
- With affordability being a key issue, many prospective homeowners are priced out of the market. Co-ownership has the potential by bringing investors to the

---

10 "What Could Work for Future Workplaces, Beyond Working?" 2018 (Hood, C., Nanda, A.) *Corporate Real Estate Journal* 7: 4352–368. https://www.ingentaconnect.com/content/hsp/crej/2018/00000007/00000004/art00006?crawler=true

cash-strapped homeowners, similar to a start-up looking for investment. Baum (2017) notes examples of sites who combine crowdfunding-sourced equity and secondary markets for residential real estate owner/occupiers and co-investors.
- Digital mortgage lending is another area of potential investment with online lenders facilitating the mortgage application process and servicing of the loan.

However, the presence of traditional banks and lending institutions is deep-rooted in residential real estate. It is difficult to imagine a demise of mortgage financing models currently in place due to the sheer size and complexities which are also backed by decades of regulatory development. However, there is no reason some of the aspects of mortgage processes (such as application, evaluation, approval and servicing) are becoming more and more technology-driven with less and less human interference and error-free. The tech companies and start-ups can work along with the traditional lending market players to create a more efficient and functional mortgage market eco-system. This can potentially alleviate three major risks in mortgage market: default and foreclosure risk (for the lender) through more accurate evaluation and predictive analytics; prepayment risk (for the lender) by finding alternative channels and investment; and equity and interest rate risk (for the borrower) by providing alternative matching facilities, market projections and valuation. The benefits are significant: (a) better matching of risks and return, (b) controlled speculation, (c) effective capital recycling and (d) tighter bandwidth for market fluctuation.

In this chapter, I have discussed the issues for the residential real estate as an asset class. I have touched on the subject of technology in the property sector. All types of technology and associated innovations are changing the urban environment. The human species is becoming more and more reliant on technology. While the benefits of technology is huge, there can be costs. In the housing context, the issues are quite significant as we increasingly depend on technology solutions for housing services. In the next and final chapter, I discuss these issues within the urban environment with specific reference to housing contexts.

Selected research topics

- Drivers of student housing and impact on university sector.
- Changes in attitude towards student housing; educational models and student housing.
- Policies and regulations for foreign investment in housing market.
- Impact of investment in digital technology on housing affordability.
- Linkages between workplace and residence in digitalised urban living.

10 Housing within sustainable, intelligent places

Chapter outline

- Market for sustainable home
- Smart city or intelligent place
- Urban operating system
- Digital services; internet of housing things
- Home as a hub; healthcare at home
- Big data in housing
- Digital inclusion and social housing
- Selected research topics

In the last section of Chapter 9, I made reference to the uses of technology in the housing market and discussed several areas of application. In this chapter, I continue with putting technology as an enabler within various habitable places and housing considerations playing a big role in it. While some of the discussions is directly about the cities and places and their challenges, housing being one of the most important consideration for daily lives and being a consumption as well as investment good plays a central role within the cities.

One of the most important arguments I make is the need to prioritise people in all technology interventions. It is crucial that wellbeing, comfort and livelihoods of all sections of our society become the most important consideration when evaluating suitability and applicability of any technological innovations. Otherwise, we risk social cohesion, leaving out a section and possibly fuelling the risk of creating inequality, which we already have plenty of, especially in the housing market. Technology can of course help alleviate or eradicate some of the banes of our times such as climate change, food shortage, potable water shortage, inadequate supply of shelter for all, lack of access to quality healthcare for all, lack of sanitation, inadequate waste management system, etc. The cost of technology is an issue, but the good news is with progress in science, the costs have come down, although it does require application of sound economic principles to price technology and protect the under-privileged and create safeguards for misuse and abuse.

Cities and urban areas, as explained in Chapter 5, face massive challenges. Due to competing economic forces applied by many stakeholders, urban areas can become a hotbed of competition for resources. Being the centre of economic activities and thus, employment, urban areas attract a population influx. Over the history of our civilisation, we have seen great cities at the forefront of creating conducive environments for innovation and societal progress. All areas of innovations that can improve the quality of human life often originate and prosper within urban areas such as science, technology, arts and humanities. At the same time, for this to happen, appropriate governance of resources through a justified set of rules and standard is key. Over the history of our times, we have also experienced lapse of governance, misuse and abuse of power leading to significant loss and costs of human society. In short, urban areas are the spatial focal points of human civilisation and need to be managed and looked after. As mentioned elsewhere, as we are approaching a time where almost two-thirds of the human population will be living in urban areas and cities, managing the need of this urbanised population is becoming a huge challenge for city or urban area administration. This is a massive challenge as the human population may reach a staggering almost 10 billion by 2050. To manage the resource use that can benefit all citizens is a stupendous task. Technological innovation can and perhaps is the only way to deal with this challenge.

At the city or urban area level, the challenges are many and require ground-level actions to tackle those. There are many opinions on the biggest challenges for the cities or urban areas. In the following discussion, I mention a list of those challenges and indicate the issues and technology solutions. I will primarily focus on new digital technology. It is important to note that the challenges are very much inter-linked and therefore the solutions need to be holistic. In all these discussions, one important point to note is that there is an equally great need for appropriate attitude towards technology, otherwise all technology solutions can easily fail. It is important to understand and analyse the user experience for all technology platforms.

Key challenges faced by urban areas

1 **Population growth and demographic diversity**

Current projection indicate that a majority of almost 10 billion people will live in cities by 2050.

There are many lists available for the largest cities in the world. While many of them provide similar answers, the geographic definition may differ and as a result, the population number also may show a difference. Regardless of those differences, the world now has many large cities. According to the estimates from the World Population Review[1], there are 123 cities with a population of more than 1 million people. Table 10.1 shows the ones with a population of 10 million or more and there are 30 of those as of 2018.

One of the alarming aspects of Table 10.1 is that except for Osaka, all these 10 million+ population cities show positive growth. And many of these are growing at a fast rate – for example, Lahore and Bengaluru grew at more than 4% from 2017; Delhi, Dhaka, Chongqing and Lagos grew at more than 3%.

Within the population growth, the demographic pattern is of huge importance. Key questions for the city authorities are: how many of these people are elderly? How many live below poverty line? How many people are employed? How many people are vulnerable or in need of social care? And many other questions. The large number of people make the challenges look much bigger. Any solution needs to incorporate the scale of issues. A vast majority of these top 30 cities are in the developing world.

Application of technology:

One of the biggest challenges of the public authorities is to determine the identity and location of the residents. These are basic problems – the primary steps are knowing who and where the residents are to understand the scale of any other problem and determine the ability to respond with a solution. Technology platforms are available for this purpose. For example, the 'Aadhaar' programme in India, which is a 12–digit

---

1 http://worldpopulationreview.com/about/

Table 10.1 Largest cities in terms of population, 2018

Name	2018 Population	2017 Population	Change
Tokyo	37,468,302	37,397,437	0.19%
Delhi	28,513,682	27,602,257	3.30%
Shanghai	25,582,138	24,862,075	2.90%
Sao Paulo	21,650,181	21,391,624	1.21%
Mexico City	21,580,827	21,500,251	0.37%
Cairo	20,076,002	19,648,312	2.18%
Mumbai	19,979,955	19,756,049	1.13%
Beijing	19,617,963	19,210,643	2.12%
Dhaka	19,578,421	18,894,385	3.62%
Osaka	19,281,188	19,289,029	−0.04%
Karachi	15,400,223	15,020,931	2.53%
Buenos Aires	14,966,530	14,879,100	0.59%
Chongqing	14,837,823	14,332,185	3.53%
Istanbul	14,750,771	14,539,767	1.45%
Kolkata	14,680,613	14,594,123	0.59%
Manila	13,482,468	13,271,721	1.59%
Lagos	13,463,421	13,042,316	3.23%
Rio De Janeiro	13,293,172	13,174,768	0.90%
Tianjin	13,214,790	13,040,664	1.34%
Guangzhou	12,638,305	12,315,664	2.62%
Moscow	12,409,738	12,288,465	0.99%
Shenzhen	11,907,836	11,693,175	1.84%
Lahore	11,738,186	11,263,275	4.22%
Bengaluru	11,440,030	10,989,540	4.10%
Paris	10,900,952	10,844,847	0.52%
Bogota	10,574,409	10,277,518	2.89%
Jakarta	10,516,927	10,401,144	1.11%
Chennai	10,455,606	10,189,373	2.61%
Lima	10,390,607	10,194,477	1.92%
Bangkok	10,156,316	9,898,653	2.60%

Source: Based on data from World Population Review, accessed on 26.08.2018

unique identity number for the residents of India, is based on their biometric and demographic data. The data is collected by the Unique Identification Authority of India (UIDAI). Aadhaar has the world's largest biometric ID system. Other countries also have similar systems. The ability to hold this massive information in digital form and be able to access it online by the authorities provides a huge advantage of understanding issues and resident needs.

2 **Insufficient and inefficient public health provision**

Linked to the first issue, population growth, as mentioned in Chapter 7, the population is ageing. Some cities have a larger young cohort but many do have a large number of elderly residents. Access to quality and affordable healthcare (hospital beds, emergency medical response) requires a massive amount of resources.

Application of technology:

A digital health system with rapid information flow can have a huge advantage. Digitised health records can help with the authorities' ability to deliver care at home (thereby reducing the need for hospital visits and hospitalisation), ability to respond to emergency needs quicker or ability to manage mass outbreak of disease, etc. These can be facilitated by technology such as mobile platforms, connected devices (wearable, mobile phone) and integrated information systems (connecting various custodians of data and patient's records). There are AI-based applications than can lower the response time and increase the efficiency and accuracy substantially. With faster internet and 4G/5G mobile connections, some of these solutions can be made viable.

3 **Food and water insecurity**

Rapid urbanisation can lead to reduction in arable land near the urban areas. With rising demand for food, supply will need to come from elsewhere, which can raise cost and also the risk of a steady supply. Many large cities periodically suffer from a water shortage. The groundwater level is declining across many cities, which creates a shortage for drinking water and other daily uses.

Application of technology:

Technology can provide effective solutions for measuring per capita water usage, extent of water wastage, monitoring water health, determining accurately periodic water needs to optimise supply, predicting water shortage for planning purposes, stopping urban water pollution, etc. through a network of sensors, digital meters, monitoring drones and other surveillance tools.

4 **Climate change, urban pollution, energy shortage**

It is now widely accepted that climate change is real. Warmer temperatures are becoming more common. Human activities in the urban areas are contributing heavily towards the massive irreversible damage to the environment and irregular weather patterns. The pollution level is rising with a poor quality of urban air and environment. This can also cause adverse health outcomes for the residents. One of the key needs for urban areas is uninterrupted energy supply. Frequent power cuts can make management of urban areas difficult. At the same time, reliance on fossil fuel-based energy needs to be heavily restricted or even eliminated. There is a great need for investing in the production of clean energy.

There is a well-established understanding that the built environment is one of the primary sources of the carbon footprint. Buildings of all kinds use a huge amount of natural resources and consume energy (electricity, gas) for daily function. Housing occupies a big part of the built environment. Being durable, houses that are built many decades ago exist today with functional uses. The addition of new housing units to the existing stock is quite low. In the developed countries, much of the housing stock is quite old. As a result, those tend to be highly energy inefficient as those concerns were not of importance during the development and was not given any significance. Insulation, use of certain materials and design features that can enhance natural lighting were not given much importance. However, with environmental concerns, this has become a key concern. With most countries pledging to reduce carbon emissions by a certain amount within acertain time period due to global warming concerns, there are now regulations for the new and existing buildings to achieve a certain level of energy efficiency. In the non-residential real estate market, there are several green building and energy efficiency standards and benchmarks such as LEED and BREAM. These certifications provide rating and performance measures on energy efficiency parameters.

In the residential real estate market, several regulations have been enacted across the countries to ensure achievement of energy efficiency standards. For example, in the UK, an energy performance certificate (EPC) is mandatory for any transaction. The EPC is a rating for the housing unit – A being the best and G being the worst. However, regulation can only work if it can be made effective in the market in a way that economic agents can see economic benefits. A crucial question is often asked in the academic and policymaking community: do people pay for more energy efficient housing units? It is reasonable to expect that there will be a significant willingness to pay for energy efficiency by housing consumers. Several studies have looked at this question. In Chapter 4, I have discussed the hedonic estimation method in detail, which is what is often used in evaluating how much a consumer would be willing to pay for a higher level of energy efficiency. If there is a price premium, we should see much better adoption and adherence to the mandatory requirement. Not just the ownership market, there is now regulations in the private renting sector (PRS) sector. The Minimum Energy Efficiency Standard (MEES), which became effective in England and Wales on 1 April 2018, is applicable to the PRS sector and non-domestic property. MEES is designed to mandate landlords to improve the energy efficiency of their PRS units. New tenancies cannot be struck for properties with an EPC rating of F and G. A similar stipulation is also applied to the existing tenancies.

Many studies have found a positive relationship between environmental labels and prices (sales and rent) in the commercial real estate (for example, Wiley et al., 2010; Eichholtz et al., 2010, 2011; Fuerst and McAllister, 2011a, 2011b; Reichardt et al., 2011; Deng et al., 2012). In the residential sector, one of the earlier studies, Brounen and Kok (2011), evaluated the relationship between EPC ratings and the sale price for 31,993 residential sale prices in 2008–9 in the Netherlands. Compared to D-rated homes, the authors find that there are significant premiums of almost 10%, 5.5% and 2% for A, B and C ratings respectively. For dwellings rated E, F and G, they estimated negative premiums or discounts of 0.5%, 2.5% and 5% respectively. In England, Fuerst et al. (2015) found similar strong effects – dwellings in EPC bands A and B sell for a 5% premium, all else equal; for C, it is 1.8%. The effect is about £8900 on an average house price (£177,824 as of August 2014). Authors also note that there is considerable variation in these effects by region and property types. In a similar study using housing transaction data for Wales, Fuerst et al. (2016) also found strong support for energy efficiency premiums: statistically significant positive price premiums for EPC bands A/B (12.8%) and C (3.5%) compared to houses in band D. For dwellings in band E (−3.6%) and F (−6.5%), there are statistically significant discounts.

Application of technology:

There is now a huge push towards developing smart environment monitoring systems. The ability to predict natural disasters such as floods, droughts and earthquakes can be enhanced by technology through sensors, AI-based analysis of data, etc. Smart metering of energy and water use can provide accurate estimates for the households. Sophisticated modelling using big data can predict peak energy demand. Mobile monitoring devices and systems for surface-level air pollution, indoor air quality, homes with intelligent appliances that can optimise energy use by domestic appliances are some of the examples of technology solutions.

5 **Traffic congestion, lack of mass transport**

Transport is perhaps the biggest function of any urban authority. With expanding urban areas and rising population density, the demand for transport can be huge. With greater use of cars and other types of personal vehicles, the need for roads has gone up. Due to environmental concerns and the fact that not all residents can afford to have personal vehicles, public transport is in much greater need than before. Mass transit systems need to be affordable, reliable and safe. But this requires massive upfront investment, which some developing countries may find it difficult to garner. Nonetheless, this is a crucial need. Poor transport systems lead to loss of economic resources and productivity and moreover it can lower urban wellbeing and lead to poor mental and physical health.

Application of technology:

Digital technology solutions can be quite effective in dealing with urban transport issues. Many cities around the world now have mass transit systems with multiple modes of transport integrated. This requires intelligent transport management systems. Network of sensors, CCTVs, linked satellite feeds, number plate tracking system, online real-time status updates and dashboards are some of the examples of technology solutions. Daily transport needs and snags can be analysed by big data analytics and the response can be modelled through machine learning frameworks.

6 **Inadequate, unaffordable housing**

Rising population implies rising demand for habitable homes. Lack of supply of housing is now a chronic problem to all cities and urban areas. Inadequate housing is leading to overconsumption with impaired personal and familial wellbeing, rising house prices leading to worsening unaffordability along with rising homelessness. Speculative activities and opportunistic development are on the rise across many cities. The housing injustice can create a concentration of urban poor such as in slums and shanty towns, social tension, inequality and a rise of illegal practices. Poor housing outcomes can significantly contribute towards many other problems such as mental and physical health, educational attainment, employment outcomes, etc.

Application of technology:

In chapter 9, I have discussed several areas of technology intervention in the housing context – from 3D printed units or sections to predictive analysis of housing demand and buyer-seller matching platforms.

7 **Poor urban sanitation**

Across all cities, including those in developed countries, urban sanitation is a big challenge. Rising population puts a huge strain on the ability to manage human waste. It can have harmful effects on residents' health, dire consequence to the environment and on the economy. Each year millions of people, most of them children and vulnerable people, suffer from diseases associated with poor sanitation and hygiene. Many people still lack basic sanitation facilities. Poor sanitation can affect women and girls more than others. Lack of safe toilets is a risk to health problems and personal safety. This is a critical need for millions of citizens.

Application of technology:

Safe management of human waste in an efficient sanitation service chain, effective resource use and recovery can be delivered by technology solutions. The solutions need to be affordable and easily accessible. For example, the Indian government has embarked on a massive campaign and programme for urban sanitation called Swachh Bharat. A massive scale of toilet construction is being rolled out around the country. The Indian city of Pune is aiming to become a Smart Sanitation City with innovative sanitation solutions supported by technology such as advanced smart waste management systems, smart technologies in the toilets and sanitation system and digital communication systems.

8 **Lack of education and barriers to human capital development and social mobility**

Education is a basic human right. Yet, a large section of the urban population lack access to quality education at all levels. Younger cohorts especially need to be able to access quality education. Otherwise, it can create huge barriers to human capital development, which is a key national asset and leads to future skill shortages and lower productivity. When cities are competing for attracting global talent and capital, lack of proper education can significantly dent the competition advantage. Movement of labour crucially depends on the quality of local education. The repercussion can also be felt on social mobility over the long-run and can be dangerous for a city's future prospect. Education and the talent pool can attract businesses which further improve employment and income outcome for the residents.

Application of technology:

Many technology solutions have come about providing quality education at all levels. With rising internet penetration and mobile technology, there are EdTech (Education Technology) platforms which can offer friction-free on-demand distance learning, bespoke training modules, access to online library resources, etc. One of the key advantages of technology solutions in education is affordability. Quality education and resources can be made available at low costs, which can especially benefit low-income residents and facilitate in social mobility through talent development.

9 **Social exclusion and economic inequality**

It is quite apt to quote (Coriolanus Act 3, Scene 1) "What is the city but the people?" People are place-shapers and place-makers. One of the key questions is: to what extent do all social classes benefit from a technological impulse to change their urban fabric? How can we prevent creating any form of social exclusion or unwanted urban divide? Social, cultural and relational capital are very important assets for a place. Social inclusion of various resident groups within public services need to be ensured to avoid fuelling urban anxiety. All social classes should benefit from information and communications technology (ICT) interventions to prevent social polarisation. A community should learn, adapt and innovate to allow for long-term sustainability and wellbeing (Coe et al., 2001). There is a crucial role of ICT along with 'soft infrastructure' such as knowledge networks, voluntary organisations and crime-free environments (Nijkamp, 2008). 'Without real engagement and willingness to collaborate and cooperate between public institutions, private sector, voluntary organisations, schools and citizens there is no smart city' (Nam and Pardo, 2011: p. 283). Nam and Pardo (2011) rightly point out that the notion of an intelligent city can only emerge at the intersection of the knowledge society (a society in which knowledge and creativity have great emphasis and intangible, human and social capital are considered the most valuable asset) with the digital city.

Application of technology:

Along with hard technology, as discussed under other challenges above, there is a crucial need to invest in soft technology that can bolster the social fabric of the urban area. Use of social technology to help residents connect and communicate is important. Open, fear-free and safe communication channels are important for ensuring social cohesion. Digital connectivity can facilitate such a process quite effectively. However, it is also important to ensure security, safety and preventions of ill-motivated uses of social technology.

10 **Lack of urban resilience**

Urban resilience can be defined as the collective ability and capacity of individual residents, various communities and all organisations (public, private and voluntary) of an urban area to withstand, change, adapt, and prosper in the face of any challenges. 100 Resilient Cities is an innovative project and organisation which is helping cities around the world to become more resilient.[2] The organisation notes that building urban resilience requires looking at a city holistically: understanding the systems that make up the city and the interdependencies and risks they may face. By strengthening the underlying fabric of a city and better understanding the potential shocks and stresses it may face, a city can improve its development trajectory and the well-being of its citizens.

It lists several challenges that may test a city's urban resilience, namely 'chronic stresses' (slow moving disasters that weaken the fabric of a city such as high unemployment, overtaxed or inefficient public transportation systems, endemic violence, chronic food and water shortages) and 'acute shocks' (sudden, sharp events that threaten a city, such as earthquakes, floods, disease outbreaks and terrorist attacks). For this purpose, the organisation notes that all systems and processes of an urban area need to be reflective (using previous experience to inform a decision, i.e. adaptive learning), resourceful (recognition of alternative uses of resources), robust (well-conceived, constructed, managed systems), redundant (keeping spare capacity to accommodate any disruption), inclusive (creating sense of shared responsibility and ownership), integrated (bringing disparate systems and institutions together) and flexible (willingness and ability to adopt alternatives for changing scenarios). Social inclusion can encourage diverse populations to come together and start collaborating and taking responsibility for their actions towards a higher level of wellbeing for all residents in the community. Incentives does not necessarily need to be monetary; intangible and non-monetary rewards and incentives can be quite productive.

Application of technology:

Social technology platforms can provide tools and platforms to deliver programmes that can lead to better understanding and ways and means of addressing the problems by the residents, which is a necessary condition for urban resilience. Checks and measures that consider the resilience aspects can help design more effective projects, programmes and policies. This is what 100 Resilient Cities calls 'the resilience dividend', i.e. 'net social, economic and physical benefits achieved when designing initiatives and projects in a forward looking, risk aware, inclusive and integrated way'.

11 **Fiscal deficit, unbalanced budget**

As we have seen in Table 10.1, many of the largest cities are in the developing world. Urban areas around the world are now facing huge budget challenges. With growing needs and rising complexities of functions, the funding requirement is huge. Many cities often face the problem of an unbalanced budget and are becoming debt-ridden. This further hampers their ability to raise funding in the globally competitive capital market. National and state governments often need to intervene and provide funding, which may not be adequate.

Application of technology:

Innovative investment matching tools, budget tools and financial applications can at least provide better understanding of cities' financial aspects and abilities. FinTech applications can be effective in delivering clear and granular financial planning and scenario preparedness.

12 **Lack of public safety, risks and fear of terrorism**

While cities and urban areas are great engines of generating economic growth, the security challenges are becoming enormous as urban areas expand with a population

---

2 https://www.100resilientcities.org/resources/

increase and expansion of urban boundaries. Not only is the expansion that matters in this context but also complexities of the security issues are on the rise. For example, as mentioned before, lack of suitable housing can lead to the spread of slums, ghettos and shanty towns along with ramifications for delivery of healthcare, transport, and water and energy infrastructure. The strain to social cohesion, rising urban anxiety along with poor urban health (mental and physical health), economic downturn etc can significantly fuel urban violence and the opportunity for anti-social elements to infiltrate and infect the urban social fabric. In the face of such issues, the public safety can be seriously compromised, which can scare away residents, talents and investments. These can only make urban areas vulnerable to terrorist attack but also make those a spawning ground for terrorist sentiment and activities.

Application of technology:

Digital tech-driven surveillance and monitoring systems (such as CCTVs, drones, sensors) along with intelligent crime analytics based on big data can provide solutions. However, only these alone cannot provide an effective solution. These need to be accompanied by programmes and soft infrastructure to create social resilience.

13 **Inadequate technology, technology risks and lack of data safety**

I have discussed many areas of application of digital technology above. With rapid deployment of digital technologies to become intelligent or smart cities, urban areas are now generating huge amounts of data. These data can be related to the residents (e.g. biometric, addresses etc), urban assets and infrastructures (e.g. healthcare, transport, emergency services, electricity grid), connected buildings and domestic devices, etc. Internet of Things (IoT) technologies such as sensors and cameras collect and continuously transmit data from the physical world. However, all these data can easily be vulnerable to cyber-attacks and crimes. Public and city systems are known targets of the hacking community. As those are connected to private citizens, the impact can be devastating. The ability to infiltrate and contaminate the urban systems is a big risk.

Application of technology:

Along with technology solutions, urban areas and cities invest much in cyber and data security. It requires careful planning and a steady injection of funds to ensure and maintain the highest level of cyber security. As technologies are becoming sophisticated, the cyber-crime entities are also becoming sophisticated. Therefore, steady investment is needed to ensure a city's preparedness towards this very significant danger.

For public authorities, one perennial problem has been estimating the needs to be able to determine supply. With a much faster computing ability and the generation of massive amounts of data, it is now possible to have a better estimation of demand and supply shifters. While some challenges are easy to track and take action on, some demand-supply shifters require new forms of data and modelling for better forecasting and assessment.

Cities ranked on key parameters

The challenges that I have discussed often become ranking parameters of cities. Ranks are important as top cities such as capital cities, global cities etc compete for attracting investments, businesses and talent. Several organisations publish a periodic ranking of top cities in terms of some of the challenges. In this section, I discuss a number of those ranks. For details

Table 10.2 Sustainable cities 2016 – ranked 1–25 and 76–100

1	Zurich	76	Riyadh
2	Singapore	77	Istanbul
3	Stockholm	78	Guangzhou
4	Vienna	79	Sao Paulo
5	London	80	Buenos Aires
6	Frankfurt	81	Jeddah
7	Seoul	82	Rio de Janeiro
8	Hamburg	83	Lima
9	Prague	84	Mexico City
10	Munich	85	Tianjin
11	Amsterdam	86	Amman
12	Geneva	87	Hanoi
13	Edinburgh	88	Jakarta
14	Copenhagen	89	Chennai
15	Paris	90	Johannesburg
16	Hong Kong	91	Bengaluru
17	Berlin	92	Mumbai
18	Rotterdam	93	Chengdu
19	Canberra	94	Wuhan
20	Madrid	95	Cape Town
21	Sydney	96	Manila
22	Rome	97	New Delhi
23	Vancouver	98	Nairobi
24	Barcelona	99	Cairo
25	Manchester	100	Kolkata

Source: Based on data from Arcadis; for a full list, refer to the source.

on individual ranks, readers are encouraged to refer to the sources for a more detailed understanding of the ranking and the underlying methodologies. These rankings not only provide a picture of the competitive landscape but also these can be the basis for devising strategies and policies for maintaining or improving on certain aspects.

The Arcadis Sustainable Cities Index ranks 100 global cities on three dimensions of sustainability: people, planet and profit, or the three 'P's.[3] These are supposed to measure performance on social, environmental and economic sustainability.

The index reports huge variation in the sustainability parameter. Compared with Table 10.1, the largest cities in terms of population are not the most sustainable ones – Tokyo is ranked 45th and New Delhi is ranked 97th in terms of sustainability (see Table 10.2). The variance in median ages across the 100 cities is huge – ranging from just 18.7 years in Nairobi to 46.6 years in Tokyo. These variations in demographic dividend is a very significant competitive advantage. For economic and social development, demographic patterns play a big role. The cities in the developed world are more likely to achieve a higher sustainability score. Two main reasons could be: smaller population size and ability to fund a programme that can improve sustainability.

Arcadis also reports individual ranks on people, planet and profit elements of the overall index.

As we can see from Tables 10.3 and 10.4, the ranks change drastically when cities are considered in terms of how they are performing on people and planet elements. For example, the

---

3 https://www.arcadis.com/en/global/our-perspectives/sustainable-cities-index-2016/#

Table 10.3 Top 100 sustainable cities 2016: People – ranked 1–25 and 76–100

1	Seoul	76	Chennai
2	Rotterdam	77	New York
3	Hamburg	78	Kolkata
4	Vienna	79	Houston
5	Berlin	80	Baltimore
6	Prague	81	Dallas
7	Amsterdam	82	Hong Kong
8	Munich	83	Indianapolis
9	Muscat	84	Tampa
10	Montreal	85	Santiago
11	Antwerp	86	Mumbai
12	Brussels	87	New Delhi
13	Barcelona	88	Buenos Aires
14	Stockholm	89	Miami
15	Frankfurt	90	Istanbul
16	Canberra	91	Lima
17	Warsaw	92	Cairo
18	Paris	93	New Orleans
19	Lyon	94	Manila
20	Madrid	95	Rio de Janeiro
21	Brisbane	96	Mexico City
22	Melbourne	97	Sao Paulo
23	Vancouver	98	Nairobi
24	Copenhagen	99	Johannesburg
25	Sydney	100	Cape Town

Source: Based on data from Arcadis; for a full list, refer to the source.

Table 10.4 Top 100 sustainable cities 2016: Planet – ranked 1–25 and 76–100

1	Zurich	76	Hanoi
2	Stockholm	77	Tianjin
3	Geneva	78	Atlanta
4	Vienna	79	Bangkok
5	Frankfurt	80	Guangzhou
6	Wellington	81	Jeddah
7	Rome	82	Riyadh
8	Sydney	83	Nairobi
9	London	84	Kuala Lumpur
10	Hamburg	85	Jakarta
11	Madrid	86	Manila
12	Singapore	87	Moscow
13	Copenhagen	88	Muscat
14	Manchester	89	Kuwait City
15	Birmingham	90	New Delhi
16	Berlin	91	Shanghai
17	Rotterdam	92	Chengdu
18	Vancouver	93	Cairo
19	Amsterdam	94	Lima
20	Glasgow	95	Abu Dhabi
21	Leeds	96	Dubai
22	Edinburgh	97	Beijing
23	Barcelona	98	Doha
24	Munich	99	Wuhan
25	Canberra	100	Kolkata

Source: Based on data from Arcadis; for a full list, refer to the source.

Table 10.5 Top 100 sustainable cities 2016: Profit – ranked 1–25 and 76–100

1	Singapore	76	Santiago
2	Hong Kong	77	Shanghai
3	London	78	Guangzhou
4	Dubai	79	Riyadh
5	Zurich	80	Cape Town
6	Edinburgh	81	Jeddah
7	Prague	82	Buenos Aires
8	New York	83	Mexico City
9	Paris	84	Sao Paulo
10	Stockholm	85	Muscat
11	Munich	86	Rio de Janeiro
12	San Francisco	87	Tianjin
13	Abu Dhabi	88	Jakarta
14	Vienna	89	Wuhan
15	Macau	90	Nairobi
16	Amsterdam	91	Manila
17	Copenhagen	92	Chengdu
18	Kuala Lumpur	93	Mumbai
19	Seoul	94	Hanoi
20	Canberra	95	Chennai
21	Washington	96	New Delhi
22	Boston	97	Amman
23	Frankfurt	98	Cairo
24	Denver	99	Bengaluru
25	Hamburg	100	Kolkata

Source: Based on data from Arcadis; for a full list, refer to the source.

top ranked city of Seoul in terms of the people element is only 26th rank in terms of planet, i.e. in terms of how green it is, abundance of green space and the quality of air, drinking water and sanitation. Many cities in the developing world (in India, China and Africa) do not rank highly on these parameters.

An interesting element that Arcadis ranking looks at is profit, i.e. economic performance (Table 10.5). The profit sub-index is related to a cities' wealth. It is measured by the city's gross domestic product (GDP) per capita. Also included are the indicators of tourism and importance to global networks mapping economic and commercial links with other cities of the world. At the same time, importance of ease of doing business and the quality of transport links matter. Cities without a good transport link and having difficulty for doing business can suffer from the lack of investment, especially in the international context. Singapore and Hong Kong top the list followed by London, Dubai and Zurich.

The Economist Intelligence Unit (EIU) produces a global liveability index.[4] The liveability index evaluates the locations in terms of providing the best or the worst living conditions. The assessment is performed on a broad range of uses and quantifies the challenges that might be presented to an individual's lifestyle in a location. Every city is assigned a rating of relative comfort for over 30 qualitative and quantitative factors across five broad categories: stability, healthcare, culture and environment, education, and infrastructure. Each factor is rated as acceptable, tolerable, uncomfortable, undesirable or intolerable. For qualitative indicators, a rating is based on the judgement of in-house analysts and in-city contributors.

---

4 http://www.eiu.com/Handlers/WhitepaperHandler.ashx?fi=The_Global_Liveability_index_2018.pdf&mode=wp&campaignid=Liveability2018

Table 10.6 The global liveability index 2018

Country	City	Rank	Overall rating	Stability	Healthcare	Culture and environment	Education	Infrastructure
Austria	Vienna	1	99.1	100	100	96.3	100	100
Australia	Melbourne	2	98.4	95	100	98.6	100	100
Japan	Osaka	3	97.7	100	100	93.5	100	96.4
Canada	Calgary	4	97.5	100	100	90	100	100
Australia	Sydney	5	97.4	95	100	94.4	100	100
Syria	Damascus	140	30.7	20	29.2	40.5	33.3	32.1
Bangladesh	Dhaka	139	38	50	29.2	40.5	41.7	26.8
Nigeria	Lagos	138	38.5	20	37.5	53.5	33.3	46.4
Pakistan	Karachi	137	40.9	20	45.8	38.7	66.7	51.8
Papua New Guinea	Port Moresby	136	41	30	37.5	47	50	46.4

Source: Based on data from The Economist Intelligence Unit; for a full list, refer to the source.

Table 10.7 The European digital city index 2016

1	London		21	Lyon
2	Stockholm		22	Aarhus
3	Amsterdam		23	Birmingham
4	Helsinki		24	Lisbon
5	Paris		25	Frankfurt
6	Berlin		26	Eindhoven
7	Copenhagen		27	Utrecht
8	Dublin		28	Cologne
9	Barcelona		29	Malmo
10	Vienna		30	Uppsala
11	Munich		31	Toulouse
12	Cambridge		32	The Hague
13	Bristol		33	Budapest
14	Madrid		34	Gothenburg
15	Oxford		35	Luxembourg
16	Manchester		36	Glasgow
17	Brussels		37	Prague
18	Tallinn		38	Warsaw
19	Edinburgh		39	Karlsruhe
20	Hamburg		40	Cardiff

Source: Based on data from Nesta; for a full list, refer to the source https://digitalcityindex.eu/.

For quantitative indicators, a rating is calculated based on the relative performance of a number of external data points. The scores are then compiled and weighted to provide a score of 1–100, where 1 is considered intolerable and 100 is considered ideal. The liveability rating is provided both as an overall score and as a score for each category.

Table 10.6 shows the best and worst cities in terms of global liveability index in 2018. Vienna and Melbourne top the list in 2018 while Damascus, Dhaka and Lagos are the bottom three in the list of 140 cities. As the EIU report notes, the rankings of cities like Damascus, Karachi and Tripoli are low and suggest that conflicts can be responsible for the low scores.

> This is not only because stability indicators have the highest single scores but also because factors defining stability can spread to have an adverse effect on other categories. For example, conflict will not just cause disruption in its own right, it will also

Housing within sustainable places 211

damage infrastructure, overburden hospitals and undermine the availability of goods, services and recreational activities. Unavailability of adequate infrastructure is also responsible for many of the lowest scores. This is particularly visible in the ranks of cities like Dhaka (Bangladesh, 139th), Harare (Zimbabwe, 135th), Douala (Cameroon, 133rd) and Dakar (Senegal, 131st).

With increasing digitalisation, cities are becoming hotbeds of innovation and connectivity plays an important role. The European Digital City Index, published by Nesta, describes how well different cities across Europe support digital entrepreneurs (refer to Table 10.7). For startups and scaleups, the index provides information about the strengths and weaknesses of local ecosystems. For policymakers, the index can be a useful tool for evaluating the need for resource spending and devise work programmes to support improvement.

Issues and challenges for creating an intelligent place

Throughout the history of human civilisation, cities and urban areas have acted as prime locations of innovation and human enterprise. With the modern day cities, this is even more important. Some of the biggest cities in the world can boast of having more residents than some of the countries on the planet and having more GDP than some of the countries. However, with increasing scale and complexities, the cities need to manage their resources and processes better to be able to provide a sustainable place for the residents. The successes and declines of cities crucially depend on this aspect.

Innovation in digital technology and massive, widespread use by millions of residents have made the city operation more challenging and complex. Starting in the 1980s with widespread use of personal computers at affordable prices, there have been efforts and initiatives for using smart technologies to help with and improve delivery of all citizen services. We can categorise those in three phases:

- Phase 1: 1980–1990 – this decade experienced early use of Information and Communication Technologies (ICT) to make cities manage their resources and perform their routine tasks otherwise handled manually by human. For example, limited use of e-transaction facilities, billing, complex budget calculations.
- Phase 2: 2000–2010 – this decade saw widespread use and penetration of the internet, which opened many channels for implementation of ICTs in city infrastructure such as e-governance and e-learning with the internet as an information exchange medium. Online transactions and customer service systems brought significant efficiency and resource gains for the cities. Online platforms became common for residents to interact and perform routine tasks related to citizen services.
- Phase 3: Present decade (2010–2020): With smartphone usage rising exponentially, earlier in this decade, the applications of wireless networks and mobile technologies (e.g. Sensors, Beacons, Social media feedback) have become popular and cost-effective Internet of Things (IoT)-based products and services are becoming increasingly widespread. Connectivity is becoming a key resource and cities are spending a huge amount of resources in achieving higher level of and more reliable connectivity.
- Future: Post-2020: It is quite clear that there will be increasing uses of AI, machine learning, robotics, autonomous vehicles and units on most citizen services.

Much of these drives are coming on the back of several key trends that are fundamentally challenging and shaping up human civilisation. These are:

- *'Humanity on the move'* – largest displacement of populations since World War II; migrant crisis: cities are primary recipients of the migrant population. As discussed in Chapter 7, being the centre of economic activities, investment and employment generation, migrants (within country as well as international) flock to cities for settling.
- *'Humanity becoming urbanised'* – with rapid urbanisation across all parts of the world, shelter, work, infrastructure, amenities etc are going through huge changes and implications of these changes are huge. Cities are at the forefront.
- *'Humanity living longer'* – medical innovation and health awareness are making humans live longer than before. Cities need to prepare for this with better management and delivery of appropriate healthcare and social care.
- *'Humanity facing climate change'* – climate change is real and affecting cities. Cities need to address the implications for resource use and disaster management with appropriate adoption and mitigation strategies.

The last couple of decades have seen huge interests in smart city initiatives. Currently, there are numerous smart city projects undertaken with huge private and public investments. The developing world is catching up quite fast with much more investment in creating smarter environments and processes to manage cities. For example, India has embarked on the 100 smart city programme.

With digitalisation, customers or residents are becoming more tech-savvy and service-aware. Demand for fast and reliable services is the order of the day. Not just in the developed countries, but also in the developing world, residents look for value for money and quality services. However, public authorities are faced with constraints and one of the main constraints is budget. As a result of that, public authorities are looking to lower costs and raise efficiency without causing any disruption of the urban public amenities and services. However, it is not just about ensuring better transactions and services. Smart city initiatives should be viewed as a city's backbone and empowering residents to participate in shaping up the city's future. Taking part in such a cause can really bring all talents together and create resilient societies. However, many smart city projects often tend to be simple injections of technology, without significant considerations for residents' user experience. Understanding the urban user experience need to be information-driven

There are several considerations that are key for successful implementation of smarter approaches to housing the urban residents and management of urban systems as identified by the literature.

- Citizen Experience (CX): As mentioned above, resident empowerment should be the key goal above and beyond other efficiency gain goals of any smart city programme. Citizen experience is crucial for this objective. If there is not a defined need from the citizens' perspective, then it is most likely a redundant investment and most like destined to fail to achieve any meaningful and significant outcome. Economising on replicating activities in this regard can bring in much needed cost and resource savings. A clear framework and mapping exercise are needed for all programmes.
- Data Awareness (DA): Data is the key resource. While systems can be put in place for security of the data, proper and purposeful use cases are required to reap the benefits of data generation, collection and storage. Huge investments in data gathering and

storing exercises may not yield much if the use cases are loosely constructed or not well-thought. For developing beneficial use cases that can add value to the citizen experience and improve the quality of urban life, data awareness by all stakeholders is a key requirement. At all levels, regardless of skillsets, functions and importance, an understanding of basic aspects of data has to be ensured. All stakeholders should ask four simple questions in this regard: (a) What data do I possess? (b) What data items do I generate from daily activities? (c) What data items should I keep and not keep? (d) How can these data items that I possess and generate be productively used? Answers to these questions do not require any a priori knowledge of mathematics, statistics, computer or data science. It simply requires an intuitive understanding of data needs and uses. With IoT-enabled management systems, products and services, data-aware stakeholders including residents can make innovation effective and beneficial.

- Digital Infrastructure (DI): Cities are making huge investments in creating a digital infrastructure that can support complex information flow. Key to this is to ensure the highest standard of security and ease of accessibility. Obviously, both are difficult to achieve and require implementation of robust open data standards. Open systems are important, as those can enable friction-free, easy to use access to key on-demand information by the residents as well as can support service innovations. Digital infrastructure should aim to connect all relevant functions through a framework that can allow function-specific operational excellence and innovation as well as cross-cutting innovation.
- Dynamic Open Feedback Loop (DOFL): Mutual understanding, wiliness to take responsibility of one's own actions and willingness to adapt and learn from adverse situations build urban resilience. For that to be achieved, openness and objectivity towards feedback is key. Mistakes, errors and omissions should be viewed as opportunities to learn and help each other. It is about enabling, empowering and encouraging the residents to provide fearless, objective feedbacks and encouraging service providers to fearlessly recognise the problem and act on the feedback. It should be two-way traffic with no frictions. However, in all these aspects, it is also important not to be willing to change to future needs and emerging trends. Revisiting all processes frequently for updates and amendments to face the future can make the whole system resilient to any future eventualities. Hard and soft physical and social infrastructure and urban assets need to be looked after from this perspective.

In view of the above discussion, let us have look at the definitions of smart city and other related terminology. Terminology and definitions are mixed and vague – several terms are used interchangeably. Table 10.8 shows several definitions of a smart city. It should be noted that citizen centric focus seems to be a key differentiator.

Figure 10.1 shows that there are many aspects of a smart city. In order to build a framework, one needs to map out various functions and sub-systems. The provision of public services is also rapidly changing. City authorities increasingly need to provide more responsive and effective services. Urbanisation and the movement of people are driving huge demand for public services. As such, the city administrations need to reduce operating budgets, cost items, resource uses and wasteful activities. Expectations and aspirations on the part of local communities are also becoming more complex. To address the new challenges, a shift in design of systems, platforms and infrastructure services is of paramount importance. These should be seamlessly linked towards a more citizen centric delivery model where informed feedback is gained from a wider range of resident demographics. Citizen centricity is a key

Table 10.8 Definition of smart city

Schuler (2002)	**Virtual city** concentrates on digital representations and manifestation of cities
Couclelis (2004)	The **digital city** is a comprehensive, web-based representation, or reproduction, of several aspects or functions of a specific real city, open to non-experts. The digital city has several dimensions: social, cultural, political, ideological, and also theoretical.
Komninos (2006)	**Intelligent cities** are territories with a high capability for learning and innovation, which is built-in the creativity of their population, their institutions of knowledge creation, and their digital infrastructure for communication and knowledge management.
Hollands (2008)	**Wired cities** refers literally to the laying down of cable and connectivity not itself necessary smart.
Anthopoulos and Fitsilis (2010)	**Ubiquitous city** (U-City) is a further extension of digital city concept. This definition evolved to the ubiquitous city: a city or region with ubiquitous information technology.
Anthopoulos and Fitsilis (2010)	Digital environments collecting official and unofficial information from local communities and delivering it to the public via web portals are called **information cities**.

Source: Based on Cocchia, A. (2014) "Smart and Digital City: A Systematic Literature Review" Dameri, R.P. and Rosenthal-Sabroux, C. (eds.), Smart City, Progress in IS

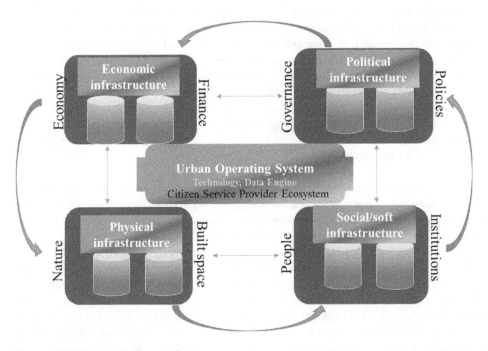

Figure 10.1 Framework of smart city

ingredient for promoting democratic functioning of cities. Trust, inclusion and empowerment can influence and strengthen the democratic processes.

In this model of public service delivery, the role of businesses is crucial and central. Private sector participation is part of the solution and endogenous to the system. This is important from the sustainability of business perspectives. A clearer understanding of a citizen's wants and needs as informed by the citizen herself can provide a very clear mandate to the business operation. Much uncertainty and costs involved with obsolescence and loss of market can be avoided by such consumer-led citizen engagement models. Fernández-Güell et al. (2016) provide a holistic framework of various functions and roles within a city framework.

Therefore, a workable definition of an 'intelligent place' should use technologies as enablers by promoting sustainable, purposeful and people-focused uses. It should focus on humanising data-led solutions to enhance the quality of daily lives of its users. Citizen engagement and co-ownership of assets and actions should be recognised as key and programmes should be devised on multi-disciplinary approaches engaging with a multi-stakeholder eco-system for effective human-centric solutions. In terms of size, it can range from a single building or a rural area or a large city region.[5]

There are several issues linked with intelligent place work programmes. A successful programme should be built on a technically sound use case that can be backed up by business and investment cases.

Issue 1: Innovation meets investment

An intelligent place embodies a market for an *intelligent* built environment and real estate, including housing. The key questions: How do we value 'built space' considering the virtual as well as physical uses of such space and how can we best leverage physical-digital value-added?

Digitalisation is happening at a much faster pace than anything else that we have seen in the history of human civilisation. Depth and breadth of this phenomenon are staggering, and implications are profound for all industries. For real estate, it is a very significant factor – it can change and reshape several traditional aspects of doing business in real estate, across all sub-sectors of the industry. This comes in at a time when internationalisation is intense too. On one side, digitalisation relates to the occupier market and building specific issues. On another side, it can affect valuation, usage and changes in the capital value of assets. In the last two decades, the real estate market has experienced an important shift towards a global platform delivery model with a network of real estate investment organisations and their support service providers. Digitalisation can influence this process further in a significant manner and in some aspects it can fundamentally change the transaction processes as discussed before. While it has some similarity to what we have experienced in terms of green building/sustainability/responsible investment in real estate, digitalisation can reach much beyond than that.

Real estate as an asset class exhibits several unique characteristics (as discussed in Chapter 2) that play key roles in determining the demand and supply, the volume of investments, the level of consumption, the profile of the investable stock and consequently, rental

---

5 Centre for intelligent Places, Henley Business School, the University of Reading, UK. https://www.henley.ac.uk/research-hub-centre-for-intelligent-places

and capital values. In addition, real estate (especially commercial real estate) is characterised by infrequent transactions and lumpy investments. All these give rise to very formidable challenges to test postulates of economic and finance theories. It can be fairly confidently argued that digitalisation has the potential to influence each of these attributes in a significant manner.

In a recent industry study by Charles Russell Speechlys, LLP, a range of issues and challenges have been highlighted for the real estate industry in the face of digitalisation.[6] As noted in the report, forward-thinking market players are recognising that a smarter built environment and intelligent real estate can deliver gains stretching far beyond just energy efficiency and sustainability. It can actually deliver on the prospect of cost savings, it can create new sources of additional revenue streams and it can profoundly change the occupiers' needs by enhancing employee productivity and improving wellbeing. In many ways, intelligent real estate can be a critical source of competitive advantage for an organisation, whether it is a real estate company or not. However, there are considerations to be made in terms of regulatory frameworks and strategic business needs. Several key possibilities can be observed in these emerging dynamics as noted in the report by Charles Russell Speechlys, LLP:

- Due to the strong ability to capture deeper level of data on usage and various cost metrics, digitalised real estate assets can lead to significant cost savings. Deeper level of data capture will enable occupiers to optimise space utilisation.
- Due to significant impact on the cost savings, it can potentially change the valuation of the smart real estate assets. Value per unit may change. The key question is – would it require a rethinking of how we value properties?
- Due to a finer level of information availability, information itself may generate revenue, further influencing the valuation of physical space with newer options. Services (e.g. health, insurance, retail, legal etc) that are delivered through built space can also experience a change in scope and opportunities.
- Real estate provisions increasingly require an eco-system of providers. Digitalisation can make it more intense. Collaboration with technology providers is becoming more and more important.

Figure 10.2 shows five critical aspects of effective place innovation – omni-connected, real-time, interoperable, deep analysis and investable. Across these attributes, productivity, inclusivity and resiliency are the key considerations.

Issue 2: Business model for *'intelligent* place' projects

Business model for intelligent projects has to work. As Figure 10.2 suggests, for an effective place innovation, one of the key requirements is 'investability'. Exploitable opportunities including working with public sector providers can ensure repeatable businesses where third-party suppliers can create a market place for products and services. Platform businesses need to be built on 'symbiotic' relationships where one party's actions and profits can also generate benefits for others. Such a 'symbiotic business model' (SBM) keeps the whole eco-system alive and functional in the long-run. Private public enterprise partnerships can drive new

---

6 http://www.charlesrussellspeechlys.com/media/312439/infographic-spread.pdf

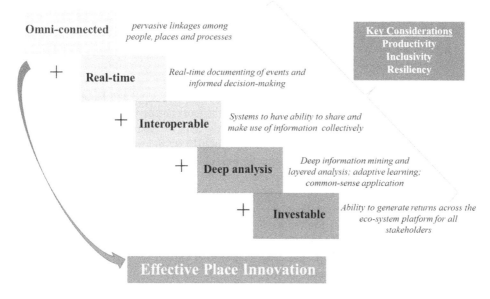

Figure 10.2 Requirement of effective place innovation.

technology products and services across the value chain, where utilisation of open shared data can act as catalysts and bring about a reduction in reliance on the state.

Three requirements for an SBM:

- Identify investment needs for *intelligent* place projects
- Induct stakeholders into the investment model
- Introduce new funding mechanisms

These should be in view of three targets of (a) reduce opex, (b) add value – improve public services; generate positive externalities and (c) generate revenue – new/enhanced revenue channels for private sector.

However, such endeavours must answer their inter-related questions: (a) who pays for the costs? (b) what are the channels of revenues? (c) who and how to share the revenues across the value chain?

Issue 3: Urban big data and complex urban operating system

To be able to provide efficient citizen services, the city administrations across the world want:

- to understand the network of existing infrastructures in their city such as water, telecom, transport, energy.
- to measure the usages, traffic and their evolution, the quality of services
- to be informed of what is happening in real-time.
- to interact with third-party operators as soon as the needs arise.

These objectives can be dealt with big and complex data being generated in the urban environment. As noted before, there is a huge amount of data being generated in the urban environment. Not only are the activities and data items linked to the physical space but also users' interactions in the virtual space from numerous digital devices located with the urban areas are creating massive amounts of data. These data items collectively form what we can call 'urban big data'. Layers and layers of disparate data items can be 'stacked' to form the universe of 'urban big data'.

The new 2018 Global Digital suite of reports from 'We Are Social' and Hootsuite reveals that there are now more than 4 billion people around the world using the internet.[7] More than half of the world's population is online, with the latest data revealing that nearly a quarter of a billion new users came online for the first time in 2017. Different parts of the world are experiencing varying speeds of internet penetration and usage. Africa has seen the fastest growth rates, with the number of internet users across the continent rising by more than 20% year-on-year. The availability of cheaper smartphone handsets and affordable tariff plans have been significant drivers for such an increase in usage. According to the report published in 2018, more than 200 million people got their first mobile device in 2017, and two-thirds of the world's 7.6 billion inhabitants have a mobile phone.

Social media use has gone up, aided by smartphone capabilities and powerful social media engines along with many apps that are developed to facilitate various forms of social interactions. The report notes that more than 3 billion people around the world now use social media each month, with 9 in 10 of those users accessing their chosen platforms via mobile devices. The number of internet users in 2018 is 4.021 billion, up 7% year-on-year. The number of mobile phone users in 2018 is 5.135 billion, up 4% year-on-year. This is also fuelling consumer markets. Worldwide, the number of people using e-commerce platforms to buy consumer goods (e.g. fashion, food, electronics and toys) grew by 8%, with almost 1.8 billion people around the world now buying goods online. Roughly 45% of all internet users now use e-commerce sites. The key question in the face of such a digital presence is: how can these digital activities shape the urban space use? That is a big question. Let's have a look at a miniaturised example. Figure 10.3 shows how various data items from a mobile network and transport system can provide improvement and efficiency gains in certain aspects of urban services.

The illustration in Figure 10.3 shows that the operating system of urban areas are becoming complex with many functions requiring integration and synchronised operational platforms. These platforms need an operating engine to be able to process multivariate inputs and deliver a desirable outcome. The complexity of the core of this Urban Operating System (UOS) requires system dynamics frameworks by integrating many functional sub-systems into one system of systems with an automatic and accurate path determination. All these require AI-enabled frameworks and much of these path detections can be performed by machine learning with no human interference. The intelligent and connected devices would be instructed to interact in a way the information churn becomes friction-free and robust.

Figure 10.4 shows a view of the UOS. Tech-enabled services and capabilities are possible when various functions (which typically exist as 'silos' in traditional city administration) are integrated and connected with seamless information sharing.

---

7 Source: https://wearesocial.com/uk/blog/2018/01/global-digital-report-2018

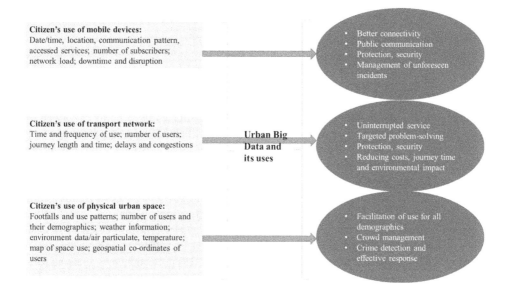

Figure 10.3 Urban big data.

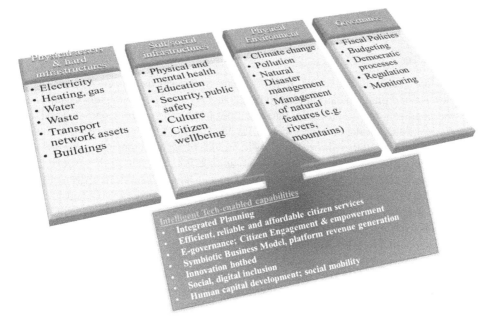

Figure 10.4 Urban Operating System (UOS).
Source: Adapted from Nasscom.

Issue 4: Intelligent place risks

One of the key requirements of making a successful intelligent place is place vision and leadership – a clear and consistent vision of the future and how that vision can be led and delivered. The vision should be co-developed with the citizens and all private and public stakeholders so that everybody shares the co-ownership and takes responsibilities of devising actions and be accountable for such actions. A culture of entrepreneurship and leadership within and across territorial units can lead the vision for better living experiences based on shared values. It also requires appropriate place branding, which is the process of image communication to a target market. As places are competing for talent, resources, and businesses with millions of small cities, thousands of large cities and hundreds of metropolises, identification of unique character and appropriate place branding are very important.

Leadership is not only important for change management but also for devising mitigation strategies for risks. There can be several risks: technology risk, operational risk, construction risk, market risk and policy risk. Time inconsistency issues can also pose a significant risk to all strategies due to dynamic factors and changing landscape. Risk management requires the regulatory support with the goals of devising Intellectual Property (IP) protection laws, creating legal frameworks to encourage innovation and building effective government support systems. The regulatory support and frameworks can protect people, data privacy and technical knowledge.

As mentioned before, one of the key motivations for intelligent urban systems is to be able to provide healthcare to the residents. In the next section, I discuss the issues and framework for providing healthcare at home and other services under the 'digital housing' concept.

Digital housing

Our homes are becoming connected. Many appliances, devices and functions have digital capabilities. From digital, connected television sets and personal computing devices to smart meters and intelligent refrigerators, the functionalities of housing as an 'economic good' can be transformed. Much of the housing services can be enhanced through these digital technology interventions. And all these have the ability to provide better and more affordable services. This is especially true when we need to manage multiple housing units such as social housing development, PRS units, retirement housing, student housing, serviced apartments, etc. Technology can deliver significant cost and functional efficiencies in these large housing operations. There is now growing adoption of digital services within these set-ups. One of the key goals of this is to establish and strengthen the communication between landlords and tenants.

Some of the digital housing services include:

1 Online billing, rent collection and other rental services
2 Online communication channels for addressing complaints, service requests and maintenance. Social media (such as Facebook, Twitter and other messaging platforms – WhatsApp, WeChat) is used for communication. Often these can include online chat services with the housing service delivery staff.
3 Feedback collection and processing: housing associations and landlords can use online platforms for rolling out surveys and other feedback collection mechanisms (such as online discussion forums).

It is also possible to use digital technology for tenants' human capital development and improving personal wellbeing. Platforms can be targeted for skill development, adult training programmes and monitoring anti-social activities, which can help in devising social programmes to address the issues. However, digital services can often suffer from lack of use and low adoption rates. Attitude towards technology use often determine the adoption rate. Therefore, it is crucial to assess the tenants' attitude and behaviours towards technology to be able to put in place an effective digital housing service programme and associated platforms. This will safeguard the costly investment.

There are several challenges that may exist for starting digital housing services. Internet penetration, poor broadband strength and affordability are key concerns. Not only across developing countries but also there are pockets of low or non-existent internet adoption in developed countries, especially in rural areas. Moreover, even if there is internet availability, low adoption rate depends on tenants' demographic pattern. Digital skill level may not be advanced enough for technology platforms to be optimally used. Much investment and attention needs to incorporate digital skill development and training programmes for the user groups. This may also include staff members who typically deliver the housing services. Having all stakeholders in the eco-system on board is important, which can be called 'inclusive digitalisation'. While poor digital skill and capabilities can pose a significant challenge, another big challenge is data security and privacy. Tenants can be justifiably concerned about the security of their private data and possibility such data to be used against their wellbeing. Regardless, inclusive digitalisation within the housing service context (for social housing, retirement housing, student housing) can deliver massive benefits:

- Improve quality of services by reducing response time and more accurate assessment of tenants' needs.
- Bring in cost efficiency and ability to pass on the cost gains to the tenants.
- Improve communication within the community; social cohesion; creating circles of support.
- Channelise resources for human capital development and social mobility.
- Lower negative events such as crimes.

Home as a hub; healthcare at home

Intelligent homes are becoming the hub of digital services. These may include retail services, energy services, mobility services and also healthcare services. With severe constraints on hospital capacity, the case for creating tools and mechanisms that can provide quality healthcare services at connected homes is quite strong. As I have discussed in Chapter 7, due to a multitude of factors and scientific progress over the last 50 years, people are living longer, and needs, tastes and preferences for old age life are also transforming. A particular need of human life is shelter, which tends to become specialised as we grow older or have significant activity constraints. Shelter also influences interactions with external social, physical and built environments such as urban form, amenities and suitable infrastructure. These innately affect wellbeing. As a result, the demand for bespoke housing and at-home care and monitoring is reaching a staggering level and can only be expected to grow in the foreseeable future, even if we believe in the most conservative estimates. At the local level, the statistics indicate several challenges for the authorities, such as creating accessible infrastructure and transport, managing healthcare and emergency care expenses, providing suitable housing and so on. These are inexorably linked and their impact increases significantly with time and varies across spatial territories. The challenges could be better addressed if individuals and social institutions were more prepared for the future. This means adapting individuals' and institutions' behaviour to accept and exploit the technology advantages. Adult social care (ASC) needs are on the rise and can potentially be addressed by devising predictive analytic tools for remote at-home care and addressing issues of loneliness and isolation. It is possible to make use of data from multiple sources to analyse behaviour change through active citizenship, requiring a data oriented feedback system for both individuals and institutions. The topics of digital healthcare, patient care at home and Active and Assisted Living (AAL) are gaining prominence. However, a suitable urban design is required for delivering the benefits. An urban area that does not support elderly living (such as infrastructure supporting elderly mobility within the urban boundary) cannot see success in this regard.

Not only within homes but also the surroundings and related service delivery spatial units need to be integrated. Health and social care systems need to be integrated. Seamless data transfer and interoperability of various custodians of data must be ensured for effective delivery of healthcare at home. These require big data analysis of patients' records with other contextual urban data. AI-assisted systems can facilitate analysis of this complex system. With the ageing population affecting almost all countries in the world, the potential for such healthcare-at-home services is enormous. Many countries have started mapping out the needs and capabilities.

Figure 10.5 shows a framework for delivering care within the intelligent home setting. Stacks of disparate data items related to the person and place need to be processed within a fast and intelligent data engine to deliver the care services. This can add several benefits for care-receivers by enabling pre-emption and near real-time responses. A preventative regime with resource efficiency gains can create significant savings for taxpayer's money. However, care receivers need to adjust their behaviour and be able to take ownership of their own wellbeing.

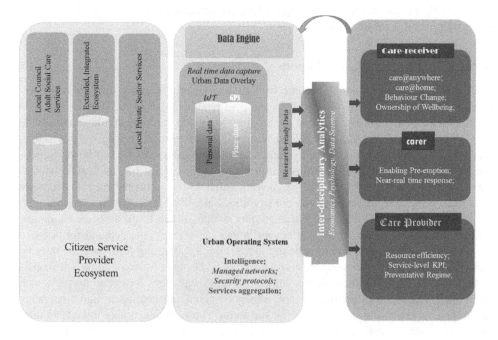

Figure 10.5 Care within intelligent home setting.

The goals of digital at-home care delivery systems should be:

- Providing a simple user interface suitable for citizens with minimal digital skills;
- Suggesting personalised programme that does not compromise on the aspirations of the citizen;
- Real-time, closed and secure feedback loop between citizens and service providers providing timely and relevant data; layers of security to ensure data privacy;
- Ability to monitor, track and benchmark performances;
- An interface to connect and share information with support networks;
- Socialisation activities analysed through dynamic interface;
- A mechanism for care and service providers' peer support networks;
- Interface for mobility services with consolidated information sharing.

A key element of delivering the above framework is appropriate skill. The healthcare delivery is no longer just about understanding the medical side of healthcare. It is also about understanding information requirement and use. Future health professionals, who may be operating in non-specialised environments (i.e. outside specialised facilities such as hospitals, nursing homes) such as homes will need to understand the eco-system and information flow and be fully trained in digital services.

Selected research topics

- What are the optimal forms of financing mechanisms for (Smart) places?
- What are the attendant risks?
- What implications does financialisation of the property market and global capital flows have for local markets and for urban areas and regions?
- What makes for a good place to grow old – safely and happily?

Bibliography

Abel, A.B. (1990). Asset prices under habit formation and catching up with the Joneses. *American Economic Review* 80(2), 38–42.
Abelson, P.W. (1979). Property prices and the value of amenities. *Journal of Environmental Economics and Management* 6, 11–28.
Abelson, P.W. and Markandya, A. (1985). The interpretation of capitalized hedonic prices in a dynamic environment. *Journal of Environmental Economics and Management* 12, 195–206.
Accetturo, A., Manaresi, F., Mocetti, S., and Olivieri, E. (2014). Don't stand so close to me: The urban impact of immigration. *Regional Science and Urban Economics* 45, 45–56.
Acemoglu, D. and Scott, A. (1994). Consumer confidence and rational expectations: Are agents' beliefs consistent with the theory? *Economic Journal* 104, 1–19.
Acemoglu, D. and Johnson, S. (2007). Disease and development: The effect of life expectancy on economic growth. *Journal of Political Economy* 115(6), 925–985.
Agnew, J. (1987). *Place and politics*. Boston: Allen and Unwin.
Aisa, R. and Pueyo, F. (2013). Population aging, health care, and growth: A comment on the effects of capital accumulation. *Journal of Population Economics* 26(4), 1285–1301.
Akbari, A. and Aydede, Y. (2012). Effects of immigration on house prices in Canada. *Appl. Econ.* 44, 1645–1658.
Akerlof, G.A. (1970). The market for 'lemons': Quality uncertainty and the market mechanism. *Quarterly Journal of Economics* 84(3), 488–500.
Akerlof, G.A. and Shiller, R.J. (2009). *Animal spirits*. Princeton: Princeton University Press.
Alexander, C. and Barrow, M. (1994). Seasonality and cointegration of regional house prices in the UK. *Urban Studies* 31, 1667–1689.
Anderson R., Lewis, D., and Zumpano, L. (1999). Residential real estate brokerage efficiency from a cost and profit perspective. *Journal of Real Estate Finance and Economics* 20(3), 295–310.
Anglin, P. M. and Arnott, R. (1991). Residential real estate brokerage as a principal-agent problem. *The Journal of Real Estate Finance and Economics* 4(2), 99–125.
Anthopoulos, L. and Fitsilis, P. (2010). From digital to ubiquitous cities: Defining a common architecture for urban development. IEEE 6th International Conference on Intelligent Environments, pp. 301–306. IEEE Xplore.
Aquilino, W.S. 1990. The likelihood of parent-child coresidence: Effects of family structure and parental characteristics. *Journal of Marriage and the Family* 52, 405–419.
Arnott, R. (1987). Economic theory and housing, in Mills, E.S. (ed.). *Handbook of regional and urban economics*. Amsterdam: (North-Holland).
Ashworth, J. and Parker, S. (1997). Modelling regional house prices in the UK. *Scottish Journal of Political Economy* 44, 225–246.
Badarinza, C. and Ramadorai, T. (2018). Home away from home? Foreign demand and London house prices. *Journal of Financial Economics* 130(3), 532–555.
Bagwell, L.S. and Bernheim, B.D. (1996). Veblen effects in a theory of conspicuous consumption. *The American Economic Review* 86(3), 349–373.

226 Bibliography

Bailey, M., Muth, R., and Nurse, H. (1963). A regression method for real estate price index construction. *Journal of the American Statistical Association* 58(304), 933–942.

Bajari, P. and Benkard, C. (2005). Demand estimation with heterogeneous consumers and unobserved product characteristics: A hedonic approach. *Journal of Political Economy* 113(6), 1239–1276.

Baker, K. and Saltes, D. (2005). Architecture billings as a leading indicator of construction. *Business Economics* 40(4), 67–73.

Baker, M. and Wurgler, J. (2006). Investor sentiment and the cross-section of stock returns. *Journal of Finance* 61, 1645–1680.

Balchin, P.N., Bull, G.H., and Kieve, J.L. (1995). *Urban land economics and public policy*, fifth edition. London: Palgrave Macmillan.

Balcilar, M., Bayene, A., and Rangan, G. (2013). 'Ripple' effects in South African house prices. *Urban Studies* 50, 876–894.

Ball, M. (1983). *Housing policy and economic power: The political economy of owner occupation*. London: Methuen & Co Ltd.

Ball, M. (2007). *Large-scale investor opportunities in residential property: An overview*. London: Investment Property Forum.

Ball, M., Blanchette, R., Nanda, A., Wyatt, P., and Yeh, J. (2011). Housing markets and independence in old age: Expanding the opportunities. http://www.mccarthyandstone.co.uk/documents/research%20 and%20policy/oorh%20full%20report%20may%202011.pdf

Ball, M.J. and Kirwan, R.M. (1977). Accessibility and supply constraints in the urban housing market. *Urban Studies* 14(1), 11–32.

Ball, M. and Nanda, A. (2013). Household attributes and the future demand for retirement housing. *International Journal of Housing Markets and Analysis* 6(1), 45–62.

Ball, M. and Nanda, A. (2014). Does infrastructure investment stimulate building supply? Case of the English regions. *Regional Studies* 48(3), 425–438.

Banks, J., Batty, G.D., Nazroo, J. and Steptoe, A (2016). The dynamics of ageing: Evidence from the English Longitudinal Study of Ageing 2002–15 (Wave 7). The Institute for Fiscal Studies. https://www.ifs.org.uk/uploads/elsa/docs_w7/ELSA%20Wave%207%20report.pdf

Bardhan A., Edelstein, R.H., and Kroll, C.A. (2012). *Global housing markets, crises, policies, and institutions*. Kolb Series in Finance. Essential Perspectives. Hoboken, NJ: John Wiley & Sons.

Bartik, T. (1987). The estimation of demand parameters in hedonic price models. *Journal of Political Economy* 95(1), 81–88.

Battiston, D., Dickens, R., Manning, A., and Wadsworth, J. (2014). Immigration and the access to social housing in the UK (CEPDP1264). Centre for Economic Performance, London School of Economics and Political Science, London, UK.

Battye, F., Bishop, P., Harris, P., Murie, A., Rowlands, R., and Rice, T. (2006). *Evaluation of key worker living*. London: DCLG.

Baum, A. (2017). PropTech 3.0: *The future of real estate*. Report, Said Business School, University of Oxford.

Bayer, P., Ferreira, F., and McMillan, R. (2007). A unified framework for measuring preferences for schools and neighbourhoods. *Journal of Political Economy* 115(4), 588–638.

Bayer, P., Ross, S.L., and Topa, G. (2008). Place of work and place of residence: Informal hiring networks and labor market outcomes. *Journal of Political Economy* 116(6), 1150–1196.

Beck, J., Scott F., and Yelowitz, A. (2012). Concentration and market structure in local real estate markets. *Real Estate Economics* 40(3), 422–460.

Beck, T., Demirgüç-Kunt, A., and Pería, M.S.M. (2011). Bank financing for SMEs: Evidence across countries and bank ownership types. *Journal of Financial Services Research* 39(1–2), 35–54.

Becker, C. M. and Morrison, A. R. (1999). Urbanization in transforming economies. Handbook of *Regional and Urban Economics* 3, 1673–1790.

Berger, A.N. and Bouwman, C.H.S. (2017). Bank liquidity creation, monetary policy, and financial crises. *Journal of Financial Stability* 30, 139–155.

Bernanke, Ben (2010). The economic outlook and monetary policy. Speech delivered at Macroeconomic Challenges: The Decade Ahead, a symposium sponsored by the Federal Reserve Bank of Kansas City, held in Jackson Hole, WY., August 26–28.

Berube, A. (2005). *Mixed communities in England: A US perspective on evidence and policy prospect.* York: Joseph Rowntree Foundation.

Black, S.E. (1999). Do better schools matter? Parental valuation of elementary education. *The Quarterly Journal of Economics* 114(2), 577–599.

Blei, D.M., Ng, A., and Michael, J.I. (2003). Latent dirichlet allocation. *Journal of Machine Learning Research* 3, 993–1022.

Blomquist, G., Berger, M., and Hoehn, J. (1988). New estimates of quality of life in urban areas. *American Economic Review* 78, 89–107.

Bloom, D.E., Börsch-Supan, A., Mcgee, P., and Seike, A. (2011). Population ageing: Facts, challenges, and responses. PGDA Working Paper No. 71. Harvard University.

Borjas, G.J. (2002). Homeownership in the immigrant population. NBER Working Paper No. 8945.

Börsch-Supan, A., Hank, K., Jürges, H., and Schröder, M. (2009). Introduction: Empirical research on health, ageing and retirement in Europe. *Journal of European Social Policy* 19, 293.

Börsch-Supan, A. and Ludwig, A. (2009). Ageing, asset markets, and asset returns: A view from Europe to Asia. *Asian Economic Policy Review* 4, 69–92.

Bourgine, P. (2004). What is cognitive economics?, in *Cognitive Economics* (pp. 1–12). Berlin, Heidelberg: Springer.

Bovaird, T. (2007). Beyond engagement and participation: User and community coproduction of public services. *Public Administration Review* 67(5), 846–860.

Bram, J. and Ludvigson, S. (1998). Does consumer confidence forecast household expenditure? A sentiment index horse race. *FRBNY Economic Policy Review*.

Bramley, G. and Dunmore, K. (1996). Shared ownership: Short-term expedient or long-term major tenure? *Housing Studies* 11(1), 105–132.

Bramley, G. and Karley, N. (2005). How much extra affordable housing is needed in England? *Housing Studies* 20(5), 685–715.

Breuer, T. (2005). Retirement migration or rather second-home tourism? German senior citizens on the Canary Islands. *Contribution to Human Geography* 136 (3), 313–333.

Brown, G.W. and Cliff, M.T. (2004). Investor sentiment and the near-term stock market. *Journal of Empirical Finance* 11, 1–27.

Brown, G.W. and Cliff, M.T. (2004). Investor sentiment and the near-term stock market. *Journal of Empirical Finance* 11, 1–27.

Burn, K. and Szoeke, C. (2016). Boomerang families and failure-to-launch: Commentary on adult children living at home. *Maturitas* 83, 9–12.

vom Brocke, J., Simons, A., Niehaves, B., Plattfaut, R., and Cleven, A. (2009). Reconstructing the giant: On the importance of rigour in documenting the literature search process. ECIS 17th European Conference on Information Systems (pp. 2–13).

Bucovetsky, S. and Glazer, A. (2014). Efficiency, equilibrium and exclusion when the poor chase the rich. *Journal of Urban Economics* 81, 166–177.

Burnley, I., Murphy, P. , and Fagan, R. (1997). *Immigration and Australian cities.* Sidney: The Federation Press.

Cannuscio, C., Block, J., and Kawachi, I. (2003). Social capital and successful aging: The role of senior housing. *Annals of Internal Medicine* 139(5_Part_2), 395–399.

Camerer, C., Loewenstein, G., and Prelec, D. (2005). Neuroeconomics: How neuroscience can inform economics. *Journal of Economic Literature* 43(1), 9–64.

Caragliu, A., Del Bo, C., and Nijkamp, P. (2011). Smart cities in Europe. *Journal of Urban Technology* 18(2), 65–82.

Carney M. (1982), Costs and pricing of home brokerage services. *Journal of the American Real Estate and Urban Economics Association* fall(3), 331–354.

Carr, P.B. and Steele, C.M. (2010). Stereotype threat affects financial decision making. *Psychological Science* 21(10), 1411–1416.

Carroll, C.D, Fuhrer, J.C., and Wilcox, D.W. (1994). Does consumer sentiment forecast household spending? If so, why? *American Economic Review* 84, 1397–1408.

Case, K. and Shiller, R. (1989). The efficiency of the market for single-family homes. *The American Economic Review* 79(1), 125-137.

Centre for Urban and Regional Development Studies. (2013). The role of social intermediaries in digital inclusion: The case of social housing. http://www.ncl.ac.uk/curds/publications/documents/RR2013-10.pdf

Chang, Y.Y., Faff, R.W., and Hwang, C-Y (2012). Local and global sentiment effects, and the role of legal, information and trading environments. SSRN. https://ssrn.com/abstract=1800550

Chari, V.V., Christiano, L., and Kehoe, P.J. (2008). Facts and myths about the financial crisis of 2008. Federal Reserve Bank of Minneapolis Working Paper, 666.

Chartered Institute of Housing Scotland. (2011). Social landlords and digital inclusion in Scotland. http://www.cih.org/resources/PDF/Scotland%20Policy%20Pdfs/Digital%20Inclusion%20Briefing%20Final.pdf

Chatterton, P. (1999). University students and city centres – The formation of exclusive geographies. The case of Bristol, UK. *Geoforum* 30, 117–133.

Cheshire, P.C. and Sheppard, S. (2004). Capitalising the value of free schools: The impact of supply characteristics and uncertainty. *The Economic Journal* 114(499), 397–424.

Chopra, N., Lee, C.M.C., Shleifer, A., and Thaler, R.H. (1993). Yes, discounts on closed-end funds are a sentiment index. *Journal of Finance* 48, 801–808.

Chourabi, H., Nam, T., Walker, S., Gil-Garcia, J. R., Mellouli, S., Nahon, K., Pardo, T. A., and Scholl, H. J. (2012). Understanding smart cities: An integrative framework. 45th Hawaii International Conference on System Sciences (pp. 2289–2297). IEEE Xplore.

Christie, H., et al. (2002). Accommodating students. *Journal of Youth Studies* 5(2), 208–235.

Chua, C.L. and Tsiaplias, S. (2009). Can consumer sentiment and its components forecast Australian GDP and consumption? *Journal of Forecasting* 28, 698–711.

Claessens, S., Kose, M.A., and Terrones, M.E. (2011). Financial cycles: What? How? When?, in *International seminar on macroeconomics* (Vol. 7, No. 1, pp. 303–344). Chicago, IL: University of Chicago Press.

Clapham, D. (2005). *The meaning of housing. A pathway approach.* Bristol: The Policy Press.

Clapham, D., Mackie, P., Orford, S., Buckley, K., and Thomas, I. with Iain Atherton and Ursula McAnulty. (2012). Housing options and solutions for young people in 2020. https://www.jrf.org.uk/sites/default/files/jrf/migrated/files/young-people-housing-options-full_0.pdf

Clapp, J.M., Nanda, A., and Ross, S.L. (2008). Which school attribute matters? The influence of school district performance and demographic composition on property values. *Journal of Urban Economics* 63(2), 451–466.

Clapp, J.M. and Ross, S.L. (2004). Schools and housing markets: An examination of school segregation and performance in Connecticut. *Economic Journal* 114, F425–F440.

Clark, W. and Dieleman, F.M. (1996). *Households and housing: Choice and outcomes in the housing market.* New Jersey: CUPR Press.

Clarke, A. (2010). Shared ownership: Does it satisfy government and household objectives, pp. 183–200 in Monk, S. and Whitehead, C. (eds.) *Making housing more affordable: The role of intermediate tenures.* Oxford: Wiley-Blackwell.

Clarke, A., Fenton, A., Markkanen, S., Monk, S., and Whitehead, C. (2008). *Understanding demographic, spatial and economic impacts on future affordable housing demand.* Cambridge: Centre for Housing and Planning Research.

Clayton, J., Ling, J., and Naranjo, A. (2009). Commercial real estate valuation: Fundamentals versus investor sentiment. *Journal of Real Estate Finance and Economics* 38(1), 5–37.

Clemens, A.W. and Axelson, L.J. (1985). The not-so-empty-nest: The return of the fledgling adult. *Family Relations* 34, 259–264.

Clotfelter, C.T. (1975). The effect of school desegregation on housing prices. *Review of Economics and Statistics* 57, 446–451.

Coase, R.H. (1937). The nature of the firm. *economica* 4(16), 386–405.

Cobb-Clark, D.A. (2008). Leaving home: What economics has to say about the living arrangements of young Australians. *Australian Economic Review* 41(2), 160–176.

Cook, S. (2003). The convergence of regional house price in the UK. *Urban Studies* 40, 2285–2294.
Cook, S. (2006) A disaggregated analysis of asymmetrical behaviour in the UK housing market. *Urban Studies* 43, 2067–2074.
Cook, S. (2012) β-convergence and the cyclical dynamics of UK regional house prices. *Urban Studies* 49, 203–218.
Cook, S. and Thomas, C. (2003). An alternative approach to examining the ripple effect in UK house prices. *Applied Economics Letters* 10, 849–851.
Cook, S. and Watson, D. (2016). A new perspective on the ripple effect in the UK housing market: Comovement, cyclical subsamples and alternative indices. *Urban Studies* 53, 3048–3062.
Cooper, H. M. (1988). Organizing knowledge syntheses: A taxonomy of literature review. *Knowledge Society* 1, 104–126.
Cosgrove, S. (2009). The United Nations framework convention on climate change. 15th Conference of the Parties – The Copenhagen Protocol. Background Paper, AMUNC.
Couclelis, H. (2004). The construction of the digital city. *Planning and Design* 31(1), 5–19 (Environment and Planning).
Coulson, N.E. (1999). Why are Hispanic-and Asian-American homeownership rates so low?: Immigration and other factors. *Journal of Urban Economics* 45(2), 209–227.
Crawford, I. and Smith, Z. (2002). Distributional aspects of inflation. *Institute for Fiscal Studies* Commentary 90, 2002.
Croce, R.M. and Haurin, D.R. (2009). Predicting turning points in the housing market. *Journal of Housing Economics* 18, 281–293.
Crockett, J.H. (1982). Competition and efficiency in transacting: The case of residential real estate brokerage. *Journal of the American Real Estate and Urban Economics Association* 10(2), 209–227.
Crook, A. and Monk, S. (2011). Planning gains, providing homes. *Housing Studies* 26(7–8), 997–1018.
Crook, A. and Whitehead, C. (2002). Social housing and planning gain: Is this an appropriate way of providing affordable housing? *Environment and Planning 'A'* 34(7), 1259–1279.
Crook, A. and Whitehead, C. (2010). Intermediate housing and the planning system, in Monk, S. and Whitehead, C. (eds.) *Making housing more affordable: The role of intermediate tenures*. Oxford: Wiley-Blackwell.
Cuaresma, J. C., Lábaj, M., and Pružinský, P. (2014). Prospective ageing and economic growth in Europe. *The Journal of the Economics of Ageing* 3, 50–57.
Cutler, D.M., Glaeser, E.L., and Vigdor, J.L. (1999). The rise and decline of the American ghetto. *Journal of Political Economy* 107, 455–506.
d'Albis, H., Boubtane, E., and Coulibaly, D. (2018). Macroeconomic evidence suggests that asylum seekers are not a 'burden' for Western European countries. *Science Advances* 4(6), eaaq0883.
Dameri, R.P. (2012). Defining an evaluation framework for digital cities implementation. IEEE International Conference on Information Society (i-Society), pp. 466–470. IEEE Xplore.
Dameri, R. P. (2013). Searching for smart city definition: A comprehensive proposal. *International Journal of Computers & Technology* 11(5), 2544–2551(Council for Innovative Research).
Das, P., Freybote, J., and Marcato, G. (2014). An investigation into sentiment-induced institutional trading behavior and asset pricing in the REIT market. *Journal of Real Estate Finance and Economics* 51(2), 160–189.
Davanzo, J. and Goldscheider, F.K. (1990). Coming home again: Returns to the parental home of young adults. *Population Studies* 44, 241–255.
Debarsy, N., Ertur, C., and LeSage, J.P. (2012). Interpreting dynamic space–time panel data models. *Statistical Methodology* 9(1), 158–171.
Degen, K. and Fischer, A.M. (2009). Immigration and Swiss house prices. CEPR Discussion Paper 7583.
Deku, S.Y. and Kara, A. (2017). Effects of securitization on banks and the financial system, securitization: Past, present and future, 10.1007/978-3-319-60128-1_5, (93–111).
Delcoure, N. and Miller, N.G. (2002). International residential real estate brokerage fees and implications for the US brokerage industry. *International Real Estate Review* 5(1), 12–39.

Della Corte, P., Sarno, L., and Sestieri, G. (2010). The predictive information content of external imbalances for exchange rate returns: How much is it worth? *Review of Economics and Statistics* Posted Online November 3, 2010. (doi:10.1162/REST_a_00157).

De Neve, J., Christakis, N., Fowler, J., and Frey, B. (2012). Genes, economics, and happiness. *Journal of Neuroscience, Psychology, and Economics* 5(4), 193–211.

DESA, United Nations: World Population Ageing 1950–2050.

Després, C. (1991). The meaning of home: Literature review and directions for future research and theoretical development. *Journal of Architectural and Planning Research* 8(2), 96–115.

Díaz, A. and Luengo-Prado, M.J. (2012). User cost, home ownership and house prices: United States. in Smith, S.J., Elsinga, M., Fox-O'Mahony, L., Ong, S.E., and Wachter, S. (eds.) *The International encyclopedia of housing and home*. Amsterdam: Elsevier.

Dieleman, F.M. (2001). Modelling residential mobility; a review of recent trends in research. *Journal of Housing and the Built Environment* 16, 249–265.

Dieleman, F.M. and Mulder, C.H. (2002). The geography of residential choice, in: Aragonés, J.I., Francescato, G., and Gärling, T. (eds.) *Residential environments: Choice, satisfaction, and behavior* (pp. 35—54). Westport, CT: Bergin and Garvey.

Di Fatta, G., Reade, J.J., Jaworska, S., and Nanda, A. (Forthcoming). Big social data and political sentiment: The tweet stream during the UK General Election 2015 Campaign. Proceedings of the 8th IEEE International Conference on Social Computing and Networking (SocialCom 2015).

DiPasquale, D. and Wheaton, W.C. (1996). *Urban economics and real estate markets*. Englewood Cliffs, NJ: Prentice-Hall.

Dominitz J. and Manski, C.F. (2003). How should we measure consumer confidence? *Journal of Economic Perspectives* 18, 51–66.

Drake, L (1995). Testing for convergence between UK regional house prices. *Regional Studies* 29, 357–366.

Dua, P. (2008). Analysis of consumers' perceptions of buying conditions for houses. *Journal of Real Estate Finance and Economics* 37, 335–350.

Duesenberry, J.S. (1949). *Income, saving, and the theory of consumer behaviour*. Cambridge, MA: Harvard University Press.

Dunleavy, P., Margetts, H., Bastow, S., and Tinkler, J. (2006). New public management is dead – Long live digital-era governance. *Journal of Public Administration* 16(3), 467–494.

Duru, D. N. and Trenz, H. J. (2016). Diversity in the virtual sphere: Social media as a platform for transnational encounters, in Sicakken, H.G. (ed.) *Integration, diversity and the making of a European public sphere* (pp. 94–115). Cheltenham: Edward Elgar.

Dykes, J. (2010). GeoVisualization and the digital city. *Computers, Environment and Urban Systems* 34, 443-451.

Easaw, J. and Heravi, S. (2004). Evaluating consumer sentiments as predictors of UK household consumption behavior: Are they accurate and useful? *International Journal of Forecasting* 20, 671–681.

Edlefsen, L. (1981). The comparative statics of hedonic price functions and other nonlinear constraints. *Econometrica* 49, 1501–1520.

Elhorst, J.P. (2012). Dynamic spatial panels: Models, methods, and inferences. *Journal of Geographical Systems* 14(1), 5–28.

Elhorst, J.P. (2014). *Spatial econometrics: From cross-sectional data to spatial panels*. Heidelberg, New York, Dordrecht and London: Springer.

Elsinga, M. (2005). Affordable and low-risk home ownership, pp. 76–94 in Boelhouwer, P., Doling, J., and Elsinga, M. (eds.) *Home ownership: Getting in, getting from, getting out*. Delft: Delft University Press.

Epple, D. (1987). Hedonic prices and implicit markets: Estimating the demand and supply functions for differentiated products. *Journal of Political Economy* 95, 59–80.

Eppright, D.R., Arguea, N.M., and Huth, W.L. (1998). Aggregate consumer expectation indexes as indicators of future consumer expenditures. *Journal of Economic Psychology* 19(2), 215–235.

Ermini, L. (1989). Some new evidence on the timing of consumption decisions and on their generating process. *Review of Economics and Statistics* 71(4), 643–650.

Ermisch, J. (1999). Prices, parents, and young people's household formation. *Journal of Urban Economics* 45, 47–71.

Estrella, A. and Mishkin, F.S. (1998). Predicting U.S. recessions: Financial variables as leading indicators. *Review of Economics and Statistics* 80(1), 45–61.

Eurocities. (2013). Cities supporting e-inclusion and citizen participation. http://nws.eurocities.eu/MediaShell/media/eInclusion%20-%20final.pdf

Fahlenbrach, R., Prilmeier, R., and Stulz, R.M. (2018). Why does fast loan growth predict poor performance for banks? *The Review of Financial Studies* 31(3), 1014–1064.

Fan, C.S., and Wong, P. (1998). Does consumer sentiment forecast ousehold spending?: The Hong Kong Case. *Economics Letters* 58(1), 77–84.

Favara, G. and Song, Z. (2014). House price dynamics with dispersed information. *Journal of Economic Theory* 149, 350–382.

Fernández-Güell, J.M., Collado-Lara, M., Guzmán-Araña, S., and Fernández-Añez, V. (2016). Incorporating a systemic and foresight approach into smart city initiatives: The case of Spanish cities *Journal of Urban Technology* 23(3), 43–67.

Fisher, J., Gatzlaff, D., Geltner, D., and Haurin, D. (2003). Controlling for the impact of variable liquidity in commercial real estate price indices. *Real Estate Economics* 31(2), 269–303.

Fisher, L. M. and Yavaş, A. (2010). A case for percentage commission contracts: The impact of a 'Race' among agents. *The Journal of Real Estate Finance and Economics* 40(1), 1–13.

Ford, J. (2006). UK Home ownership to 2010 and beyond: Risks to lenders and households. Home ownership: Getting in, getting from, getting out, Part II. *Housing and Urban Policy Studies* 30, 201–220.

Ford, J. and Wilcox, S. (1993). *Mortgage rescue*. York: Joseph Rowntree Foundation.

Forrest, R. (1987). Spatial mobility, tenure mobility, and emerging social divisions in the UK housing market. *Environment and Planning A* 19(12), 1611–1630.

Forrest, R. and Murie, A. (1988). *Selling the welfare state*. London: Routledge.

Freeman, C.E., Scafidi, B., and Sjoquist, D.L. (2005). Racial segregation in Georgian public schools, 1994–2001: Trends, causes, and impact on teacher quality, in Boger, J.C., Edley, C., and Orfield, G. (eds.) *Re-segregation of the American South*. Chapel Hill, NC: University of North Carolina Press.

Fuerst, F., McAllister, P., Nanda, A., and Wyatt, P. (2015). Does energy efficiency matter to homebuyers? An investigation of EPC ratings and transaction prices in England. *Energy Economics* 48, 145–156.

Fuerst, F., McAllister, P., Nanda, A., and Wyatt, P. (2016). Energy performance ratings and house prices in Wales: An empirical study. *Energy Policy* 92, 20–33.

Fuhrer, J.C. (1993). What role does consumer sentiment play in the U.S. macroeconomy? *Federal Reserve Bank of Boston New England Economic Review* 32–44.

Gallent, N., Mace, A., and Tewdwr-Jones, M. (2002). Delivering affordable housing through planning: Explaining variable policy usage across rural England and Wales. *Planning Practice & Research* 17(4), 465–483.

Garrod, G.D. and Willis, K.G. (1992). Valuing goods' characteristics: An application of the hedonic price method to environmental attributes. *Journal of Environmental Management* 34, 59–76.

Gartenberg, C. and Pierce, L. (2016). Subprime governance: Agency costs in vertically integrated banks and the 2008 mortgage crisis. *Strategic Management Journal* 38(2), 300–321.

Geltner, D., Kluger, B.D., and Miller, N.G. (1991). Optimal price and selling effort from the perspectives of the broker and seller. *Real Estate Economics* 19(1), 1–24.

Genesove, D. and Mayer, C. (2001). Loss aversion and seller behaviour: Evidence from the housing market. No. w8143. National Bureau of Economic Research.

George, H. (1879). *Progress and poverty*. New York: Schalkenbach Foundation.

Gibbons, S. and Machin, S. (2003). Valuing English primary schools. *Journal of Urban Economics* 53, 197–219.

Gibson, J., McKenzie, D., and Stillman, S. (2011). The impacts of international migration on remaining household members: Omnibus results from a migration lottery program. *Review of Economics and Statistics* 93(4), 1297–1318.

Gilbert, F.W., Tudor, R.K., and Paolillo, J.G.P. (1994). The decision making unit in the choice of a long term health care facility. *Journal of Applied Business Research* 10(2), 63–73.

Gibbons, S., Machin, S., and Silva, O. (2013). Valuing school quality using boundary discontinuities. *Journal of Urban Economics* 75, 15–28.

Gil de Zúñiga, H., Veenstra, A., Vraga, E., and Shah, D. (2010). Digital democracy: Reimagining pathways to political participation. *Journal of Information Technology & Politics* 7(1), 36–51.

Gilleard, C., Hyde, M., and Higgs, P. (2007). The impact of age, place, aging in place, and attachment to place on the well-being of the over 50s in England. *Research on Aging* 29(6), 590–605.

Gonzales, L. and Ortega, F. (2012). Immigration and housing booms: Evidence from Spain. *Journal of Regional Science* 53, 37–59.

Goodchild, B. and Cole, I. (2001). Social balance and mixed neighbourhoods in Britain since 1979. *Environment and Planning* D 19, 103–121.

Goodman, A.C. (1990). Demographics of individual housing demand. *Regional Science and Urban Economics* 20: 83–102.

Goodman, J.L. (1994). Using attitude data to forecast housing activity. *Journal of Real Estate Research* 9(4), 445–453.

Granger, C.W.J. (1988). Some recent developments in the concept of causality. *Journal of Econometrics* 39, 199–212.

Granger, C.W.J. and Newbold, P. (1974). Spurious regressions in econometrics. *Journal of Econometrics* 35, 143–159.

Grant, C. and Padula, M. (2016). The repayment of unsecured debt by European households. *Journal of the Royal Statistical Society: Series A (Statistics in Society)* 181(1), 59–83.

Green, R.K. (2008). Imperfect information and the housing finance crisis: A descriptive overview. *Journal of Housing Economics* 17(4), 262–271.

Gurney, C. (1999). Pride and prejudice: Discourses of normalisation in public and private accounts of homeownership *Housing Studies* 14(2), 163–183.

Gurran, N., Milligan, V., Baker, D., and Bugg, L. (2007). *International practice in planning for affordable housing: Lessons for Australia.* AHURI Positioning Paper Series, 99. Australian Housing and Urban Research Institute.

Guardian. (2016). London mayor launches unprecedented inquiry into foreign property ownership. 30 September.

Hall, E.T. (1959). *The silent language*. New York: Doubleday.

Hall, M. and Greenman, E. (2013). Housing and neighborhood quality among undocumented Mexican and Central American immigrants. *Social Science Research* 42(6), 1712–1725.

Hall, R.E. (1978). Stochastic implications of the life cycle-permanent income hypothesis: Theory and evidence. *Journal of Political Economy* 86, 971–87.

Han, L. and Hong, S.H. (2011). Testing cost inefficiency under free entry in the real estate brokerage industry. *Journal of Business and Economic Statistics* 29(4),564–578.

Hanushek, E.A., Kain, J.F., and Rivkin, S.G. (2002). New evidence about Brown V. Board of Education: The complex effects of school racial composition on achievement. Working paper # 8741, National Bureau of Economic Research.

Hauge, Å. and Kolstad, A. (2007). Dwelling as an expression of identity. A comparative study among residents in high-priced and low-priced neighbourhoods in Norway. *Housing, Theory and Society* 24(4), 272–292.

Hayek, F.A. (1942). Scientism and the study of society. *Economica* 9(35), 267–291.

Hayek, F.A. (1952). *The sensory order. An inquiry into the foundations of theoretical psychology*. London: Routledge & Kegan Paul.

Hayek, F.A. (1963). Rules, perception and intelligibility. *Proceedings of the British Academy* XLVIII, 321–344.

Heinig, S. and Nanda, A. (2018). Measuring sentiment in real estate – A comparison study. *Journal of Property Investment and Finance* 36(3), 248–258.

Helderman, A. and Mulder, C. (2007). Intergenerational transmission of homeownership: The roles of gifts and continuities in housing market characteristics. *Urban Studies* 44(2), 231–247.

Hendershott, P. and Slemrod, J. (1983). Taxes and the user cost of capital for owner-occupied housing. *Journal of the American Real Estate and Urban Economics Association* 10(4), 375–93.

Himmelberg, C., Mayer, C., and Sinai, T. (2005). Assessing high house prices: Bubbles, fundamentals and misperceptions. *Journal of Economic Perspectives* 19(4), 67–92.

Hirshleifer, D. (2001). Investor psychology and asset pricing. *Journal of Finance* 56(4), 1533–1597.

Hohenstatt, R. and Kaesbauer, M. (2014). GECO's weather forecast for the U.K. housing market: To what extent can we rely on Google Econometrics? *The Journal of Real Estate Research* 36(2), 253–282.

Hollands, R.G. (2008). Will the real smart city please stand up? *City: Analysis of Urban Trends, Culture, Theory, Policy, Action* 12(3), 303–320.

Hollifield, J. F. (2004). The emerging migration state 1. *International Migration Review* 38(3), 885–912.

Holmans, A. (1990). House price: Changes through time at national and sub-national level. Working paper 110, December. London: Government Economic Service.

Holmes, M. and Grimes, A. (2008). Is there long-run convergence among regional house prices in the UK? *Urban Studies* 45, 1531–1544.

Holmes, M.J. (2007). How convergent are regional house prices in the United Kingdom? Some new evidence from panel data unit root testing. *Journal of Economic and Social Research* 9, 1–17.

Hood, C. and Nanda, A. (2018). What could work for future workplaces, beyond working? *Corporate Real Estate Journal* 7(4), 352–368.

House of Commons Communities and Local Government Committee. (2013). The private rented sector. First Report of Session 2013–14. Located at: http://www.publications.parliament.uk/pa/cm201314/cmselect/cmcomloc/50/50.pdf

Housing Technology. (2011). Digital by default 2012. http://www.housing-technology.com/research/digital-default-2012/

Howrey, E. P. (2001). The predictive power of the index of consumer sentiment, pp. 175–207. *Brookings papers on economic activity*.

Hsieh, C. and Moretti, E. (2003). Can free entry be inefficient? Fixed commissions and social waste in the real estate industry. *Journal of Political Economy* 111(5), 1076–1122.

Huang, B. and Rutherford, R. (2007). Who you going to call? Performance of realtors and non-realtors in a MLS setting. *The Journal of Real Estate Finance and Economics* 35(1), 77–93.

Hubbard, P. (2008). Regulating the social impacts of studentification: A Loughborough case study. *Environment and Planning A* 40, 323–341.

Ihlanfeldt, K. and Boehm, T.P. (1987). Government intervention in the housing market: An empirical test of the externalities rationale. *Journal of Urban Economics* 22(3), 276–290.

Ishida, T. (2000). Understanding digital cities, pp. 7–17 in Ishida, T. and Isbister, K. (eds.). *Digital cities*. LNCS, vol. 1765. (Berlin: Springer).

Ishida, T. and Hiramotsu, K. (2001). An augmented web space for digital cities. Proceedings of Symposium on Applications and the Internet (pp. 105–112).

Jin, C., Soydemir, G., and Tidwell, A. (2014). The U.S. housing market and the pricing of risk: Fundamental analysis and market sentiment. *Journal of Real Estate Research* 36(2), 187–219.

Jordà, Ò., Schularick, M., and Taylor, A.M. (2016). The great mortgaging: Housing finance, crises and business cycles. *Economic Policy* 31(85), 107–152.

Joseph Rowntree Foundation. Housing Pathways of Immigrants. http://www.jrf.org.uk/sites/files/jrf/2103-housing-immigration-asylum.pdf

Kahneman, D. and Tversky, A. (1979). Prospect theory: An analysis of decision under risk. *Econometrica: Journal of the Econometric Society* 47(2), 263–291.

Kahneman, D. and Tversky, A. (2000). (eds.), *Choices, values and frames*. New York: Cambridge.

Kain, J. F. and Quigley, J.M. (1970). Measuring the value of housing quality. *Journal of the American Statistical Association* 65(330), 532–548.

Kain, J. F. and Quigley, J.M. (1975). Housing markets and racial discrimination: A microeconomic analysis. A selection from an out-of-print volume from the National Bureau of Economic Research, 1–8, available online: http://www.nber.org/chapters/c3711.

Katona, G. (1951). *Psychological analysis of economic behavior.* New York: McGraw-Hill.

Katona, G. (1960). *The powerful consumer: Psychological studies of the American economy.* New York: McGraw-Hill.

Katona, G. (1964). *The mass consumption society.* New York: McGraw-Hill.

Katona, G. (1975). *Psychological economics.* New York: Elsevier.

Kavarnou, D. and Nanda, A. (2015). House price dynamics in Panama City. *Journal of Real Estate Literature* 23(2), 315–334.

Kemeny, J. (1981). *The myth of home-ownership.* London: Routledge.

Kemeny J. (1995). *From public to social market: Rental policy strategies in comparative perspective.* London: Routledge.

Kim, S.H., Kim, H.B., and Gon Kim, W. (2003). Impacts of senior citizens' lifestyle on their choices of elderly housing. *Journal of Consumer Marketing* 20(3), 210–226.

Kenyon, E. L. (1997). Seasonal sub-communities: The impact of student households on residential communities. *The British Journal of Sociology* 48(2), 286–301.

Kiley, M.T. (2010). Habit persistence, nonseparability between consumption and leisure, or rule-of-thumb consumers: Which accounts for the predictability of consumption growth? *Review of Economics and Statistics* 92(3), 679–683.

Kingdon, G., Sandefur, J., and Teal, F. (2006) Labour market flexibility, wages and incomes in sub-Saharan Africa in the 1990s. *African Development Review* 18(3), 392–427.

Koblyakova, A. and White, M. (2017). Supply driven mortgage choice. *Urban Studies* 54(5), 1194–1210.

Komninos, N. (2006). The architecture of intelligent cities: Integrating human, collective and artificial intelligence to enhance knowledge and innovation. IEEE 2nd IET International Conference on Intelligent Environments (pp. 13–20). IEEE Xplore.

Komninos, N. (2008). *Intelligent cities and globalization of innovation networks.* London: Routledge.

Kumar, A. (2018). Do restrictions on home equity extraction contribute to lower mortgage defaults? Evidence from a policy discontinuity at the Texas border. *American Economic Journal: Economic Policy* 10(1), 268–297.

Kumar, A. and Lee, C.M.C. (2006). Retail investor sentiment and return comovements. *Journal of Finance* 61(5), 2451–2486.

Lauretta, E. (2018). The hidden soul of financial innovation: An agent-based modelling of home mortgage securitization and the finance-growth nexus. *Economic Modelling* 68, 51–73.

Lawhon, M. and Murphy, J. (2012). Socio-technical regimes and sustainability transitions: Insights from political ecology. *Progress in Human Geography* 36(3), 354–378.

Lea, M. (2010, February). Alternative forms of mortgage finance: What can we learn from other countries?. In *Moving Forward in Addressing Credit Market Challenges: A National Symposium.* Cambridge, MA: Harvard Business School Press.

Lean, H. and Smyth, R. (2013). Regional house prices and the ripple effect in Malaysia. *Urban Studies* 50, 895–922.

Lee, C.-C. and Chien, M.-S. (2011). Empirical modelling of regional house prices and the ripple effect. *Urban Studies* 48(10), 2029–2047.

Leeper, E.M. (1992). Consumer attitudes: King for a day. *Federal Reserve Bank of Atlanta Economic Review* 77(3), 1–16.

Leguizamon, S. (2010). The influence of reference group house size on house price. *Real Estate Economics* 38(3), 507–527.

Leguizamon, S.J. and Ross, J.M. (2012). Revealed preference for relative status: Evidence from the housing market. *Journal of Housing Economics* 21(1), 55–65.

Lewis, D. and R. Anderson. (1999). Is franchising more cost efficient? The case of the residential real estate brokerage industry. *Journal of Real Estate Economics* 27(3), 545–560.

Ley, D. and Tuchener, J. (1999). Immigration and metropolitan house prices in Canada, Research on immigration and integration in the metropolis, Working Paper No. 99-09, Vancouver Center of Excellence.

Li, M.M. and Brown, H.M. (1980). Micro-neighbourhood externalities and hedonic housing prices. *Land Economics* 56(2), 125–141.

Li., W. and Nanda, A. (2019). Housing market choices: A semiotic analytical framework. Working paper.

Linden, F. (1982). The consumer as forecaster. *The Public Opinion Quarterly* 46(3), 353–360.

Lindh, T. and Malmberg, B. (2009). European Union economic growth and the age structure of the population. *Economic Change and Restructuring* 42(3), 159–187.

Ling, D. (2009). A random walk down Main Street: Can experts predict returns on commercial real estate? *Journal of Real Estate Research* 27(2), 137–154.

Ling, D., Naranjo, A., and Scheick, B. (2013). Investor sentiment, limits to arbitrage, and private market returns. *Real Estate Economics* 41(2), 1–47.

Ling, D., Ooi, J.T.L., and Le, T.T. (2015). Explaining house price dynamics: Isolating the role of nonfundamentals. *Journal of Money, Credit, and Banking* 47, 87–125.

Liu, K. and Li, W. (2014). *Organisational semiotics for business informatics*. London and New York: Routledge.

Lucas, R. (1976). Econometric policy evaluation: A critique. Carnegie-Rochester Conference Series on Public Policy 1, 19–46.

Lund, B. (2011). *Understanding housing policy. Understanding welfare: Social issues, policy and practice*, second edition. Bristol: Policy Press.

Lux, M.(2003). *Housing policy: An end or a new beginning?* Budapest: Local Government and Public Service Reform Initiative/Open Society Institute.

MacDonald, R. and Taylor, M. (1993). Regional house prices in Britain: Long-run relationships and short-run dynamics. *Scottish Journal of Political Economy* 40, 43–55.

MacLennan, D. (1982). *Housing economics*. London and New York: Longman.

Malgarini, M. and Margani, P. (2009). Psychology, consumer sentiment and household expenditures. *Applied Economics* 39(13), 1719–1729.

Malpass, P. (2007). More coherent than sustainable: A critique of contemporary British housing policy. Paper presented at the Housing Studies Association Conference, April 2007, University of York.

Malpezzi, S. (2003). Hedonic pricing models: A selective and applied review, in Sullivan, T.O. and Gibbs, K. (eds). *Housing economics and public policy: Essays in honor of Duncan Maclennan*. Malden, MA: Blackwell.

Mankiw, G. (1982). Hall's consumption hypothesis and durable goods. *Journal of Monetary Economics* 10(3), 417–425.

Mankiw, N.G. and Weil, D. (1989). The baby boom, the baby bust, and the housing market. *Regional Science and Urban Economics* 19, 235–258.

Marcato, G. and Nanda, A. (2016). Information content and forecasting ability of sentiment indicators: Case of real estate market. *Journal of Real Estate Research* 38(2), 165–203.

Massey, D.S. (2004). Measuring undocumented migration. *International Migration Review* 38(3), 1075–1103.

Matsudaira, J.D. (2016). Economic conditions and the living arrangements of young adults: 1960 to 2011. *Journal of Population Economics* 29(1), 167–195.

Matsusaka, J.G. and Sbordone, A.M. (1995). Consumer confidence and economic fluctuations. *Economic Inquiry* 33(2), 296–318.

Mayda, A.M. (2006). Who is against immigration? A cross-country investigation of individual attitudes toward immigrants. *Rev. Econ. Stat.* 88, 510–530.

McDonald, J.F. and McMillen, D.P. (2011). *Urban economics and real estate: Theory and policy*, second edition. Wiley.

McKee, K. (2010). Promoting homeownership at the margins: The experience of low-cost homeownership purchasers in regeneration areas. *People, Place & Policy* 4(2), 38–49.

McKee, K. (2011). Challenging the norm? The 'ethopolitics' of low-cost homeownership in Scotland. *Urban Studies* 48(16), 3399–3413.

McKee, K.J., Kostela, J., and Dahlberg, L. (2014). Five years from now: Correlates of older people's expectation of future quality of life. *Research on Ageing* 37(1), 18–40.

Meen, G. (1999). Regional house prices and the ripple effect: A new interpretation. *Housing Studies* 14(6), 733–753.

Meen, G. (2002). The time-series behavior of house prices: A transatlantic divide?. *Journal of Housing Economics* 11(1), 1–23.

Meen, G. and Andrew, M. (2008). Planning for housing in the post-Barker era: Affordability, household formation and tenure choice. *Oxford Review of Economic Policy* 24(1), 79–98.

Meen, G. P. (2001). *Modelling spatial housing markets: Theory, analysis, and policy.* London: Kluwer Academic Publishers.

Metcalf, K.M., Singhvi, A., Tsalikian, E., Tansey, M.J., Zimmerman, M.B., Esliger, D.W., and Janz, K.F. (2014). Effects of moderate-to-vigorous intensity physical activity on overnight and next-day hypoglycemia in active adolescents with type 1 diabetes. *Diabetes Care* 37(5), 1272–1278.

Miceli, T.J. (1991). The multiple listing service, commission splits, and broker effort. *Real Estate Economics* 19(4), 548–566.

Miceli, T.J. (1992). The welfare effects of non-price competition among real estate brokers. *Journal of the American Real Estate and Urban Economics Association* 20(4), 519–532.

Mickelson, R.A. (2003). The academic consequences of desegregation and segregation: Evidence from the Charlotte–Mecklenburg schools. *North Carolina Law Review* 81, 1513–1562.

Milani, F. (2011). Expectation shocks and learning as drivers of the business cycle. *The Economic Journal* 121, 379–401.

Miles, D. (1994). *Housing, financial markets and the wider economy.* Chichester: John Wiley.

Miles, S. (2000). *Youth lifestyles in a changing world.* Buckingham: Open University Press.

Mill, J.S. (1984). On the definition and method of political economy. Daniel Hausman (Hrsg.), 41–58.

Miller, N.G. and Shedd, P.J. (1979). Do antitrust laws apply to the real estate brokerage industry?. *American Business Law Journal* 17(3), 313–339.

Mishkin, F.S. (1978). Consumer sentiment and spending on durable goods. *Brookings Papers on Economic Activity* 1, 217–232.

Molin, E., Oppewal, H., and Timmermans, H. (1996). Predicting consumer response to new housing: A stated choice experiment. *Netherlands Journal of Housing and the Built Environment* 11(3, Special Issue), 297–313.

Monk, S. (2009). Understanding the demand for social housing in the United Kingdom. Some implications for policy. *International Journal of Housing Markets and Analysis* 2(1), 21–38.

Monk, S., Holmans, A., Jones, M., Lister, D., Short, C., and Whitehead, C. (2007). *The demand for social rented housing: A review of data sources and supporting case study evidence.* London and Cambridge: ODPM and Cambridge Centre for Housing and Planning Research.

Mooradian, R. and Hotchkiss, E. (1997). Vulture investors and the market for control of distressed firms. *Journal of Financial Economics* 43, 401–432.

Moutinho, J.L. (2008). Building the information society in Portugal: Lessons from the digital cities program 1998–2000, in van Geenhuizen, D.M., Trzmielak, D., Gibson, D.V., and Urbaniak, M. (eds.), Value-added partnering and innovation in a changing world.

Muellbauer, J. and Murphy, A. (1997). Booms and busts in the UK housing market. *The Economic Journal* 107(445), 1701–1727.

Mulder, C.H. and Hooimeijer, P. (1999). Residential relocations in the life course, in van Wissen, L.J.G. and Dykstra, P.A. (eds.) *Population issues, an interdisciplinary focus.* New York: Kluwer Academic/Plenum Publishers.

Muller, T. and Espenshade, T.J. (1985). *The fourth wave: California's newest immigrants.* Washington: The Urban Institute Press.

Murie, A. (1997). The social rented sector, housing and the welfare state in the UK. *Housing Studies* 12(4), 437–462.

Mussa, A., Nwaogu, U.G., and Pozo, S. (2017). Immigration and housing: A spatial econometric analysis. *Journal of Housing Economics* 35, 13–25.

Muth, J.F. (1961). Rational expectations and the theory of price movements. *Econometrica: Journal of the Econometric Society* 29(3), 315–335.

Nam, T. and Pardo, T. (2011). Conceptualizing smart city with dimensions of technology, people, and institutions. In Proceedings of the 12th Annual International Digital Government Research Conference: Digital government innovation in challenging times, pp. 282–291.

Nanda, A. (2007). Examining the NAHB/Wells Fargo Housing Market Index (HMI). *Housing Economics* National Association of Home Builders, Washington, DC.

Nanda, A., Clapp, J., and Pancak, K. (2016). Do laws influence the cost of real estate brokerage services? A state fixed effects approach. *Real Estate Economics* 44(4), 918–967.

Nanda, A. and Pancak, K. (2010). Real estate brokers' duties to their clients: Why some states mandate minimum service requirements. *Cityscape: A Journal of Policy Development and Research* 12(2), 105–126.

Nanda, A. and Parker, G. (2011). Analysis of the intermediate housing market mechanism in the UK. Working Papers in Real Estate & Planning. 17/11. Working Paper. University of Reading, Reading. pp. 26.

Nanda, A. and Parker, G. (2013). Shared ownership: Easy wins but hard lessons? *Town and Country Planning* 82(4), 178–182.

Nanda, A. and Parker, G. (2015). Shared ownership and affordable housing: A political solution in search of a planning justification? *Planning Practice & Research* 30 (1), 101–113.

Nanda, A. and Ross, S. L. (2012). The impact of property condition disclosure laws on housing prices: Evidence from an event study using propensity scores. *The Journal of Real Estate Finance and Economics* 45(1), 88–109.

Nanda, A. and Tiwari, P. (2013). *Sectoral and spatial spillover effects of infrastructure investment: A case study of Bengaluru, India*. London: The Royal Institution of Chartered Surveyors (RICS).

Nanda, A. and Yeh, J. (2016). Reflected glory vs. repulsive envy: What do the Smiths feel about the house of the Joneses? *Asian Economic Journal* 30(3), 317–341.

Needham, C. (2008). Realising the potential of co-production: Negotiating improvements in public services. *Social Policy and Society* 7(2), 221–231.

Nightingale, T.E., Walhin, J.P., Thompson, D., and Bilzon, J.L.J. (2015). Influence of accelerometer type and placement on physical activity energy expenditure prediction in manual wheelchair users. *PloS One* 10(5).

Norris, D. (2002). The impact of school quality and race on neighborhood housing values. Paper presented at the AREUEA Winter Meeting.

Nutley, S. and Webb, J. (2000). Evidence and the policy process, pp. 13–41 in Davies, H., Nutley, S., and Smith, P. (eds.) *What works? Evidence based policy and practice in the public services* (Bristol: Policy Press).

Nygaard, C. (2011). International migration, housing demand and access to homeownership in the UK. *Urban Studies* 48(11), 2211–2229.

Oates, W.E. (1969). The effects of property taxes and local public spending on property values: An empirical study of tax capitalisation and the Tiebout hypothesis. *Journal of Political Economy* 77(6), 957–971.

Oates, W.E. (1973). The effects of property taxes and local public spending on property values: A reply and yet further results. *Journal of Political Economy* 81(4), 1004–1008.

OECD. (2011). M-Government. Mobile technologies for responsive government and connected societies. Technical report, OECD Publishing.

Ohtake, F. and Shintani, M. (1996). The effect of demographics on the Japanese housing market. *Regional Science and Urban Economics* 26(2), 189–201.

O'Sullivan, A. (2003). *Housing economics and public policy*. Oxford: Wiley-Blackwell.

Ott, N. (2012). *Intrafamily bargaining and household decisions*. Berlin: Springer Science & Business Media.

Pardo, T. and Taewoo, N. (2011). Conceptualizing smart city with dimensions of technology, people, and institutions. Proceedings of the 12th Annual International Conference on Digital Government Research (pp. 282–291). New York: ACM.4.

Parigi, G. and Schlitzer, G. (1997). Predicting consumption of Italian households by means of survey indicators. *International Journal of Forecasting* 13(2), 197–209.

Paris, C. (2007). International perspectives on planning and affordable housing. *Housing Studies* 22(1), 1–9.

Pastalan, L.A. (1995), *Housing decisions for the elderly: To move or not to move*. London: Routledge.

Payne, J.E. (2012). The long-run relationship among regional housing prices: An empirical analysis of the US. *Regional Analysis and Policy* 42, 28–35.

Peirce, C. S. (1902). Logic as semiotic: The theory of signs. *Philosophical writings of Peirce*, 100.

Peirce, C. S. (1958). *Collected papers of Charles Sanders Peirce. VIII: Review, correspondence, and bibliography*. Cambridge, MA: Harvard University.

Penman, S. and McNeill, L.S. (2008). Spending their way to adulthood: Consumption outside the nest. *Young Consumers* 9(3), 155–169.

Petersen, W., Holly, S., and Gaudoin, P. (2002). *Further work on an economic model of the demand for social housing*. Report number 153. London: Department of the Environment, Transport and Regions.

Phan, D. (2004). From agent-based computational economics towards cognitive economics, in *Cognitive Economics* (pp. 371–398). Berlin, Heidelberg: Springer.

Piger, J.M. (2003). Consumer confidence surveys: Do they boost forecasters' confidence? *Regional Economist*, April, Federal Reserve Bank of St. Louis (2003).

Piggott, J. (2007). *Residential transition amongst the elderly*. Sidney: Australian Institute for Population Ageing Research, University of New South Wales.

Pollakowski, H.O. (1973). The effect of the property taxes and local public spending on property values: A comment and further results. *Journal of Political Economy* 81(4), 994–1003.

Porcelli, A.J. and Delgado, M.R. (2009). Acute stress modulates risk taking in financial decision making. *Psychological Science* 20(3), 278–283.

Poterba, J. (1984). Tax subsidies to owner-occupied housing: An asset market approach. *Quarterly Journal of Economics* 99(4), 729–752.

Poterba, J. and Sinai, T. (2008). Tax expenditures for owner-occupied housing: Deductions for property taxes and mortgage interest and the exclusion of imputed rental income. *American Economic Review* 98(2), 84–89.

Poterba, J.M. (2004). Impact of population ageing on financial markets in developed countries. *FRB Kansas City – Economic Review* 89(4), 43–53.

Quigley, J.M. (2002). Transaction costs and the housing markets. Program on Housing and Urban Policy. Working Paper.

Raco, M. (2008). Key worker housing, welfare reform and the new spatial policy in England. *Regional Studies* 42, 737–751.

Reback, R. (2005). House prices and the provision of local public services: Capitalization under school choice programs. *Journal of Urban Economics* 57(2), 275–301.

Ricardo, D. (1821). *The principles of taxation and political economy*. London: JM Dent.

Ricciardi, F. (2010). ICTs in an ageing society: An overview of emerging research streams, pp. 37–44 in D'Atri, A., De Marco, M., Braccini, A.M., and Cabiddu, F. (eds.), *Management of the interconnected world*. ITAIS, vol. 1 (Berlin, Springer).

Roback, J. (1982). Wages, rents, and the quality of life. *Journal of Political Economy* 90, 1257–1278.

Robinson, D. (2010). The neighbourhood effects of new immigration. *Environment and Planning A* 42(10), 2451–2466.

Robinson, D. and Reeve, K. (2006). Neighbourhood experiences of new immigration. York: Joseph Rowntree Foundation.

Rosen, S. (1974). Hedonic prices and explicit markets: Production differentiation in pure competition. *Journal of Political Economy* 82, 34–55.

Ross, S. L. and Yinger, J. (1999). Sorting and voting: A review of the literature on urban public finance, in Cheshire, P. and Mills, E.S. (eds.) *The handbook of urban and regional economics* vol. 3, North-Holland, Amsterdam.

Rossi-Hansberg, E., Sarte, P.D., and Owens, R., III. (2010). Housing Externalities. *Journal of Political Economy* 118(3), 485–535.

Rothschild, M. and Stiglitz, J. (1976). Equilibrium in competitive insurance markets. *Quarterly Journal of Economics* 90(4), 629–649.

Rugg, J. (1999). *Young people, housing and social policy*. New York: Psychology Press.

Rugg, J., et al. (2002). Studying a niche market: UK students and the private rented sector. *Housing Studies* 17(2), 289–303.

Rutherford, R.C., Springer, T.M., and Yavas, A. (2005). Conflicts between principals and agents: Evidence from residential brokerage. *Journal of Financial Economics* 76, 627–665.

Rutherford, R.C., Springer, T.M., and Yavas, A. (2007). Evidence of information asymmetries in the market for residential condominiums. *Journal of Real Estate Finance and Economics* 35(1), 23–38.

Rykwert, J. (1991). House and home. *Social Research* 58(1), 51–62.

Sá, F. (2011). Immigration and house prices in the UK. IZA Discussion Paper 5893.

Sá, F. (2016). The effect of foreign investors on local housing markets: Evidence from the UK. CEPR Discussion Paper No. 11658.

Sabia, J.J. (2008). There's no place like home. *Research on Aging* 30(1), 3–35.

Saiz, A. (2003). Room in the kitchen for the melting pot: Immigration and rental prices. *Review of Economics and Statistics* 85(3), 502–521.

Saiz, A. (2007). Immigration and housing rents in American cities. *Journal of Urban Economics* 61(2), 345–371.

Saiz, A. and Wachter, S. (2011). Immigration and the neighborhood. *American Economic Journal: Economic Policy* 3(2), 169–188.

Sala-i-Martin, X., Doppelhofer, G., and Miller R.I. (2004). Determinants of long-term growth: Bayesian averaging of classical estimates (BACE) approach. *American Economic Review* 94, 813–35.

Santero, T. and Westerlund, N. (1996). Confidence indicators and their relationship to changes in economic activity. OECD Economics Department Working Paper No. 170.

Saunders, P. (1984). Beyond housing classes: The sociological significance of private property rights in means of consumption. *International Journal of Urban and Regional Research* 8, 202–227.

Scanlon, K. (2010). International experience, pp. 37–61 in Monk, S. and Whitehead, C. (eds.) *Making housing more affordable: The role of intermediate tenures* (Oxford: Wiley-Blackwell).

Scanlon, K. and Kochan, B. (eds.) (2011). *Towards a sustainable private rented sector: The lessons from other countries*. London: LSE.

Schaffers, H., Ratti, C., and Komninos, N. (2012). Special issue on smart applications for smart cities –New approaches to innovation: Guest editors' introduction. *Journal of Theoretical and Applied Electronic Commerce Research* 7(3), 2–10. (Universidad de Talca, Chile).

Schiewe, J., Krek, A., Peters, I., Sternberg, H., and Traub, K. P. (2008). HCU research group "Digital City": Developing and evaluating tools for urban research, in Ehlers et al. (eds.). *Digital earth summit on geoinformatics*.

Schmeling, M. (2009). Investor sentiment and stock returns: Some international evidence. *Journal of Empirical Finance* 16, 394–408.

Schuler, D. (2002). Digital cities and digital citizens, in Tanabe, M., van den Besselaar, P., and Ishida, T. (eds.). *Digital cities II: Computational and sociological approaches*. LNCS, vol. 2362 (pp. 71–85). Berlin: Springer.

Schuurman, D., Baccarne, B., De Marez, L., and Mechant, P. (2012). Smart ideas for smart cities: Investigating crowdsourcing for generating and selecting ideas for ICT innovation in a city context. *Journal of Theoretical and Applied Electronic Commerce Research* 7(3) 49–62. (Universidad de Talca, Chile).

Schwartz, A.E., Ellen, I.G., Voicu, J., and Schill, M.H. (2006). The external effects of placebased subsidized housing. *Regional Science and Urban Economics* 36, 679–707.

Scotland's digital future: A strategy for Scotland. Scottish Government (2011). http://www.gov.scot/Resource/Doc/343733/0114331.pdf

Sen, A.K. (1977). Rational fools: A critique of the behavioural foundations of economic theory. *Philosophy & Public Affairs* 6(4), 317–344.

Shaw, R.P. (1975). Migration theory and fact: A review and bibliography of current literature. https://www.popline.org/node/519619

Shiller, R. (2005). *Irrational exuberance*, second edition. Princeton, NJ: Princeton University Press.

Silverstein, M. and Zablotsky, D.L. (1996). Health and social precursors of later life retirement – community migration. *Journal of Gerontology: Social Sciences* 51B(3), S150–156.

Skifter Andersen, H., Andersson, R., Wessel, T., and Vilkama, K. (2016). The impact of housing policies and housing markets on ethnic spatial segregation: Comparing the capital cities of four Nordic welfare states. *International Journal of Housing Policy* 16(1), 1–30.

Sorrentino, M. and Simonetta, M. (2013). Incentivising inter-municipal collaboration: The Lombard experience. *Journal of Management and Governance* 17(4), 887–906.

Souleles N. (2004). Expectations, heterogenous forecast errors and consumption: Micro evidence from the Michigan Consumer Sentiment Surveys. *Journal of Money, Credit and Banking* 36, 39–72.

Spencer, I. R. (2002). *British immigration policy since 1939: The making of multi-racial Britain.* London: Routledge.

Stadelmann, D. (2010). Which factors capitalize into house prices? A Bayesian average approach. *Journal of Housing Economics* 19, 108–204.

Stamper, R. (1988). Analysing the cultural impact of a system. *International Journal of Information Management* 8(2), 107–122.

Stigler, G.J. and Sherwin, R.A. (1985). The extent of the market. *Journal of Law and Economics* 28(3), 555–585.

Stillman, S. and Maré, D.C. (2008). Housing markets and migration: Evidence from New Zealand. Working Paper 08_06, Motu Economic and Public Policy Research.

Stock J.H. and Watson, M.W. (2002a). Macroeconomic forecasting using diffusion indexes. *Journal of Business and Economic Statistics* 20, 147–162.

Stock, J.H. and Watson, M.W. (2002b). Forecasting using principal components from a large number of predictors. *Journal of the American Statistical Association* 97, 1167–1179.

Stone, J., Berrington, A., and Falkingham, J. (2011). The changing determinants of UK young adults' living arrangements. *Demographic Research* 25, 629–666.

Strassmann, W.P. (1991). Housing market interventions and mobility: An international comparison. *Urban Studies* 28(5), 759–771.

Strassmann, W.P. (2000). Mobility and affordability in US housing. *Urban Studies* 37(1), 113–126.

Strassmann, W.P. (2001) Residential mobility: Contrasting approaches in Europe and the United States. *Housing Studies* 16(1), 7–20.

Su, K., Li, J., and Fu, H. (2011). Smart city and the applications. IEEE International Conference on Electronics, Communications and Control (ICECC), pp. 1028–1031(IEEE Xplore).

Suval, E. M. and Hamilton, C. H. (1965). Some new evidence on educational selectivity in migration to and from the south. *Social Forces* 43(4), 536–547.

Sweeney, J.L. (1974). A commodity hierarchy model of the rental housing market. *Journal of Urban Economics* 1(3), 288–323.

Tang, F. and Lee, Y. (2011). Social support networks and expectations for aging in place and moving. *Research on Aging* 33(4), 444–464.

Tapscott, D. and Tapscott, A. (2016). *Blockchain revolution: How the technology behind bitcoin is changing money, business, and the world.* New York: Penguin.

Thaler, R.H. (2000). From homo economicus to homo sapiens. *Journal of Economic Perspectives* 14(1), 133–141.

Thomsen, J. (2007). Home experiences in student housing: About institutional character and temporary homes. *Journal of Youth Studies* 10(5), 577–596.

Thünen, J.V. (1826). *Der isolierte Staat. Beziehung auf Landwirtschaft und Nationalökonomie.* Hamburg: Perthes.

Tiebout, C. (1956). A pure theory of local expenditures. *Journal of Political Economy* 64(5), 416–424.

Tunstall, R. (2003). 'Mixed tenure' policy in the UK: Privatisation, pluralism or euphemism? *Housing Theory and Society* 20(3), 153–159.

Turnbull, G.K. (1996). Real estate brokers, nonprice competition and the housing market. *Real Estate Economics* 24(3), 293–316.

Universities UK. (2006). *Studentification: A guide to opportunities, challenges and practice*. London: Universities UK.

Utaka, A. (2003). Confidence and the real economy: The Japanese case. *Applied Economics* 35, 337–342.

Veblen, T. (1898). Why is economics not an evolutionary science? *Quarterly Journal of Economics* 12, 373–397.

Veblen, T. (1899). *The theory of the leisure class*. Chicago, IL: University of Chicago Press.

Verma, R. and Verma, P. (2008). Are survey forecasts of individual and institutional investor sentiments rational? *International Review of Financial Analysis* 17, 1139–1155.

Vuchelen, J. (2004). Consumer sentiment and macroeconomic forecasts. *Journal of Economic Psychology* 25, 493–506.

Wachter, S.M. (1987). *Residential real estate brokerage: Rate uniformity and moral hazard*, Research in Law and Economics series. Greenwich, CT and London: JAI Press, 10, 189–210.

Wallace, A. (2008). *Achieving mobility in the intermediate housing market: Moving up and moving on?* Report for the JRF. London: Chartered Institute of Housing.

Weber, W. and Devaney, M. (1996). Can consumer sentiment surveys forecast housing starts? *Appraisal Journal* 4, 343–350.

Weimer, D.L. and Wolkoff, M.J. (2001). School performance and housing values using non-contiguous district and incorporation boundaries to identify school effects. *National Tax Journal* 45, 231–254.

Welsh Government. (2010). Delivering digital inclusion: A strategic framework for Wales. http://gov.wales/docs/dsjlg/publications/comm/101208deliveringdien.pdf

White, L. and Edwards, J. (1990). Emptying the nest and parental well-being: Evidence from national panel data. *American Sociological Review* 55, 235–242.

Whitehead, C. (2007). Planning policies and affordable housing: England as a successful case study? *Housing Studies* 22(1), 25–44.

Whitehead, C. (2010). Shared ownership and shared equity: Reducing the risks of home-ownership? Report for the JRF. http://www.jrf.org.uk/sites/files/jrf/homeownership-risk-reduction-summary.pdf

Wilcox, D.W. (1992). The construction of U.S. consumption data: Some facts and their implications for empirical work. *American Economic Review* 82(4), 922–941.

Wilhelmsson, M. (2008). The evidence of buyer bargaining power in the Stockholm residential real estate market. *Journal of Real Estate Research* 30(4), 475–500.

Woodall, P. and Brobeck, S. (2006). State real estate regulation: Industry dominance and its consumer costs. Consumer Federation of America. http://www.consumerfed.org/elements/www.consumerfed.org/file/housing/CFA_Real_Estate_Commissioner_Report.pdf

Yavas, A. (1992). A simple search and bargaining model of real estate markets. *Journal of the American Real Estate and Urban Economics Association* 20(4), 533–548.

Yavas, A. and Colwell, P. (1999). Buyer brokerage: Incentive and efficiency implications. *Journal of Real Estate Finance and Economics* 18(3), 259–277.

Yinger, J. (1981). A search model of real estate broker behavior. *The American Economic Review* 71(4), 591–605.

Yinger, J., Bloom, H., Borsch-Supan, A., and Ladd, H. (1988). *Property taxes and house values: The theory and estimation of intra-jurisdictional property tax capitalization*. San Diego: Academic Press.

Zardini, A., Mola, L., Vom Brocke, J., and Rossignoli, C. (2010). The role of ECM and its contribution in decision-making processes. *Journal of Decision Systems* 19(4), 389–406.

Zorlu, A., Mulder, C.H., and van Gaalen, R. (2014). Ethnic disparities in the transition to home ownership. *Journal of Housing Economics* 26(4), 151–163.

Index

100 smart city programme, India 212

Aadhaar programme, India 199–200
AAL (Active and Assisted Living) 222
access to knowledge, location choice 70–1
Accetturo, A. 125
Acemoglu, D. 11, 149
Active and Assisted Living (AAL) 222
adaptive expectation 42
adult social care (ASC) 222
affordability 57, 59; housing for migrants 135–41; shared ownership 97–105
affordability gap 103
affordability ratio 138
African-American representation, house prices 82
ageing populations 146–8; housing wealth 157; isolation 152; life expectancy 153–5; retirement housing 157–8
ageing with effort 151
aggregate housing demand 31–2
aggregate supply function 33
AI (Artificial Intelligence) 191
air pollution, location choice 84
Airbnb 195
animal spirits 8, 42
ANNs (Artificial Neural Networks) 191–2
annual cost of homeownership 87–8
annuity mortgages 168
anticompetitive 113–14
AR (Augmented Reality) 192
Arcadis Sustainable Cities Index 207
ARMs (adjustable rate mortgage), United Kingdom 176
Artificial Intelligence (AI) 191
Artificial Neural Networks (ANNs) 191–2
ASC (adult social care) 222
asset classes 179; cross-border investments 186–8; PRS (private renting sector) 180–2; student housing 183–6; technology for residential real estate 188–90
attributes of: housing markets 4–10; of neighborhoods 83–4
Augmented Reality (AR) 192

Ball, M. 45, 156, 158, 160–1, 181
balloon mortgages 168
Barker, K. 77
Barker Review 77
barriers to human capital development and social mobility, urban areas 203–4
Baum, A. 194
Bayer, P. 107
Beck, T. 115, 164
betterment 79–80
BHPS (British Household Panel Survey) 142
bid rent 73–5
bid-rent function 75–6; migration 128–9
big data 190–1; intelligent places 217–19
Black, S.E. 82
blockchain technology 192, 196
Bram, J. 11
BRICS (Brazil, Russia, India, China, South Africa), GDP (Gross Domestic Product) 19–20
British Household Panel Survey (BHPS) 142
brokerage fees 111
budgets, urban areas 205
Burnley, I. 123
business models for intelligent places 216–17
buy to let mortgages 170

capital cost, location choice 67
CARA (constant absolute risk aversion) 172
case studies on migration 129–32
Case-Shiller Index 52
cashback mortgages 169
CBD (central business district) 73
CDO (collateralised debt obligation) 172
central business district (CBD) 73
challenges of, urban areas 199–206
Cheshire, P.C. 82
China: GDP (Gross Domestic Product) 19–21; student housing 183–4
cities, ranked on key parameters 206–11
citizen centricity 213–14
Citizen Experience (CX) 212
Clapham, D. 142–3, 145

Clapp, J.M. 49, 83
climate change, urban areas 201–2
CMOs (collateralised mortgage obligations) 172
Coase, R.H. 79
co-living 195
collateralised debt obligation (CDO) 172
collateralised mortgage obligations (CMOs) 172
commission rates 111–13
constant absolute risk aversion (CARA) 172
constant relative risk aversion (CRRA) 172
consumption 16
Coasian framework 79
covered bond markets 173
Crawford, R. 157
Croce, R.M. 11
cross-border investments 186–8
CRRA (constant relative risk aversion) 172
Cuba, migration 124
culture, housing and 7–14
CX (Citizen Experience) 212

Data Awareness (DA) 212–13
data centres 190
data sources for housing 37–8
debt, household debt per cent of annual disposable income 177
debt investors 179–80
deep learning 192
demand for housing 25–6; aggregate housing demand 31–2; urban areas 206; violating law of demand 26–34
demand-supply mismatch, England 137
demographics: old people 145–51; young people 141–5
DI (Digital Infrastructure) 213
Díaz, A. 85–6
Dieleman, F.M. 107
digital cities 214
digital connectivity, location choice 71–2
digital housing 220–1
Digital Infrastructure (DI) 213
digital mortgage lending 196
digitalisation 215–16
DINKOs (dual income, no kids owners) 142
direct technology application in home search 193–4
Do It Yourself Shared Ownership 102
dual agency 110
dual income, no kids owners (DINKOs) 142
durability of housing units 5
Dynamic Open Feedback Loop (DOFL) 213
dynamics of migration, key factors for consideration 133–4

ECB (European Central Bank) 175
economic agents, rational agents 8
economic inequality, urban areas 204

economic migration 119
economics, land-use planning 77–80
Economist Intelligence Unit (EIU) 209
economy, simple model of the economy 16–21
education: location choice 81–3; urban areas 203–4
EIU (Economist Intelligence Unit (EIU) 209
elasticities 28, 31
ELSA (English Longitudinal Survey of Ageing) 152–3
Elsinga, M. 98, 99–101
energy costs, location choice 67–9
energy performance certificate (EPC) 201
energy shortages, urban areas 201–2
England: intermediate housing mechanisms 101–2; real estate agencies 110; shared ownership 97–101
environment, State of the World's Cities 65
environmental migration 119
EPC (energy performance certificate) 201
equilibrium: hedonic equilibrium 44, 48; implications in housing market 34–5; market equilibrium 33
equilibrium prices 140
equity investors 179–80
estimation strategy, hedonic model 48–50
ethnic networking 127–8
European Digital City Index 210–11
evaluation aspects for homeowners in relation to housing choice 14
evolutionary economics 8
expenditure approach to GDP 18
external effects, housing markets 6

Fannie Mae 171
Federal Home Loan Mortgage Corporation (FHLMC) 171
Federal National Mortgage Association 171
fees, brokerage fees 111–13
FHLMC (Federal Home Loan Mortgage Corporation) 171
Finnish Right of Occupancy 100
FinTech 194
first time buy mortgages 170
Fisher, J. 113
fixed rate mortgages 169
food insecurity, urban areas 201
foreign investment in housing markets 186–8
Freddie Mac 171
Fuerst, F. 47, 49, 202

Gateway to Global Aging Data 153
GCI (Global Connectivity Index) 72
GDP (Gross Domestic Product) 16–25
General Data Protection Regulation (GDPR) 190
geography, housing market areas 36–7

Germany: housing financing 178; tenure choices 95
Gibbons, S. 82
Giffen goods 27
Global Connectivity Index (GCI) 72
global house prices 57–60
Global Real Estate Transparency Index 39–40
global urbanisation 61–2; megacities, 63–4
Goodman, J.L. 126
goods 26
governance, State of the World's Cities 65
government purchases 17
Gross Domestic Product (GDP) 16–25
guidelines for user cost method 89–90
Gurney, C. 101

Halifax House Price Index 50–1
Hall, R.E. 13, 127
Haurin, D.R. 11
healthcare at home 221–3
Hecht-Nielsen, R. 191
hedonic equilibrium 48
hedonic model 43–52
hedonic price function 48
Heinig, S. 12
Helsinki, migration 128
heterogeneity, housing markets 5
Himmelberg, C. 86–7
HMI (Housing Market Index) 11
Homebuy Direct 102
House Price Index 52
house price indices 41–52
house price measures 52–7
house prices 39
household bid curves 47
household debt 177
housing data sources 37–8
housing deprivation 98–9
housing financing, technology for residential real estate 196–7
housing loan penetration 165
housing market areas, sub-regional housing market areas 36–7
Housing Market Index (HMI) 11
housing markets 4–7
housing mobility 105–8
housing search process 109–11
housing tenure choices 88–97
housing transactions, search process 109–11
housing wealth, old people 157
human factors, soft human factors 9–10

ICT (information and communications technology) 204
Illinois Association of REALTORS 114
IMF (International Monetary Fund), global house prices 57

immigration 120–7
immovability, housing markets 4
impact of migrants on demand for housing 134–5
imputed rent 86–8
inadequate housing, urban areas 203
inadequate technology, urban areas 206
income approach to GDP 18
income elasticity of demand 28
India: 100 smart city programme 212; Aadhaar programme 199–200; GDP (Gross Domestic Product) 19–21
indifference curves 31
industry associations 113
information and communications technology (ICT) 204
information asymmetry, housing markets 5–6
information cities 214
infrastructure investments, role in markets 45
intelligent places 217–23
inter-bank lending 174
interest rates 168
interest-only mortgages 169
intermediate housing mechanisms, England 101–2
International Monetary Fund (IMF), global house prices 57
interpretant 13
investability 216
investments 17, 179–80; cross-border investments 186–8; PRS (private renting sector) 180–2; student housing 183–6; technology for residential real estate 188–90
isolation, ageing populations 152

JLL (Jones Lang Lasalle), Global Real Estate Transparency Index 39–40
joint sole agency contracts 110

Kavarnou, D. 51–2
KEI (Knowledge Economy Index (KEI) 71
Kemeny, J. 127

labour supply and costs, location choice 66–7
lack of urban resilience 204–5
Land Registry House Price Index 52
land rent, location choice 67–8
land-use planning 76–80
law of demand, violating 26–34
LEHC (Low Equity Housing Cooperative) 100
Ley, D. 123
Li, M.M. 13
life cycle theory 122
life expectancy 153–5
life-cycle housing model 88
Ling, D. 11
liveability index 209–10
Loan Performance Home Price Index 52

loan to value (LTV) ratios 175
local amenities, migration 125
location choice 66–8; housing and neighbourhood 80–4
locational attributes 45
locational distribution 76
longevity of housing units 5
Low Equity Housing Cooperative (LEHC) 100
LTV (loan to value) ratios 175
Ludvigson, S. 11
Luengo-Prado, M.J. 85–6

Machin, S. 82
machine learning 191
Marcato, G. 11
marginal propensity to consume (MPC) 25
marginal propensity to save (MPS) 25
marginal social benefit (MSB) 77
marginal social cost (MSC) 77
Mariel Boatlift 124, 126
market adjustments, equilibrium 34
market clearing adjustments 30
market equilibrium 27, 30, 33
mass transportation, urban areas 202–3
MBS (mortgage-backed security) 171–2, 174
McKee, K. 98
MEES (Minimum Energy Efficiency Standard) 202
megacities 63–4
mega-trends 1
migration 117–29; affordable housing 135–41; impact of migrants on demand for housing 134–5; key factors of consideration while modelling the dynamics of migration 133–4; policy concerns 132–3; spatial distribution of migrants 134; United Kingdom 129–31; United States 131–2
Miles report 176
Miller, N.G. 111
Minimum Energy Efficiency Standard (MEES) 202
minimum service laws 114–15
MLS (multiple listing service) 115
mobility, housing mobility 105–8
models of urban areas 72–6
mortgage depth 165
mortgage financing 163–6; sources of 173–8; types of 168–71
mortgage market process 166–7, 170
mortgage markets 164
mortgage tenure 168
mortgage-backed security (MBS) 171–2, 174
MPC (marginal propensity to consume) 25
MPS (marginal propensity to save) 25
MSB (marginal social benefit) 77
MSC (marginal social cost) 77
Muller, T. 123

multi agency contracts 110
multiplier effect, GDP (Gross Domestic Product) 25
Mussa, A. 125

Nam, T. 204
Nanda, A. 11, 12, 13, 36–7, 45, 50, 51–2, 59, 98, 100, 113, 156, 158, 189–90, 195
National Association of REALTORS 110
National Bureau of Economic Research (NBER) 19
National Composite House Price Index 52
The National Geographic 118
natural disasters, cities 64
Natural Language Processing (NLP) 192
NBER (National Bureau of Economics Research) 19
negative externalities 50
neighbourhoods, location choice 80–4
net exports 17
neural networks 191
New Build Homebuy 102
NIMBYism (not-in-my-backyard) 78
NLP (Natural Language Processing) 192

OECD (Organization for Economic Cooperation and Development), immigration 120
offset mortgages 170
old people: housing for 145–51; housing wealth 157
omitted variable bias 48
Open Market Home Buy 102
OORH (owner-occupied retirement housing) 156
opportunity cost of capital 86
owner-occupied retirement housing (OORH) 156
owning, versus renting 88–97

Pancak, K. 113, 114
Pardo, T. 204
Parker, G. 98, 100
PBSA (Purpose Built Student Accommodation) 184
permanent income hypothesis 43
personal traits, housing and 7–14
physical space use 194–5
Pigouvian tax 78
policies for retirement housing 162
policy concerns, migration 132–4
political migration 119
pollution, urban areas 201–2
population growth 62; immigration 120; urban areas 199–200
positive externalities 50
price determination 32
price function, hedonic price function 48
price inflation 136–8
Principal-Agent issue 112

private renting sector (PRS) 88, 92–6, 180–2
product offer curves 47
production approach, GDP 18
PropTech 189, 194
Prospect Theory 9
PRS (private renting sector) 88, 92–6, 180–2
public health provisions, urban areas 200–1
public safety, urban areas 205–6
public services, location choice 70
pull factors, migration 119–120
Purpose Built Student Accommodation (PBSA) 184
push factors, migration 119

quality, hedonic model 45

ranking parameters of cities 206–11
rational agents 8
rational expectation 42
real estate agencies, England 110
real estate brokers, United States 110–11
REALTOR 113–14
reflected glory 50
relocation, housing mobility 105–8
rent, imputed rent 86–8
Rent to Homebuy 102
renting: immigration 123–7; versus owning 88–97; social rental housing 104–6
repayment mortgages 169
repayment schemes 168
repeat sales model 51–2
repulsive envy 50
residential mortgage-backed securities (RMBS) 174
residential real estate brokerage (RREB) industry 111
resilience dividend 205
retail savings deposits 173
retirement housing 148–9, 157–8; OORH (owner-occupied retirement housing) 156; policies for 162
retirement villages 150
Ricardo, D. 73
risks: intelligent places 220; mortgage financing 168
RMBS (residential mortgage-backed securities) 174
Roback, J. 44
Robinson, D. 122, 152
rooms per person 159
Rosen, S. 43
Ross, S. L. 80, 83
RREB (residential real estate brokerage) industry 111

Sá, F. 188
Saiz, A. 123–4

sales, smart contracting of property sales 196
sanitation, urban areas 203
Savills Global Research 1
SBM (symbiotic business model) 216–17
schools, location choice 81–3
Scott, A. 11, 115
search process for housing 109–11
secondary mortgage market 171
secondary mortgage market products 172–3
segregation 81
semiotics 12
sentiment 42; housing and 7–14
shared ownership 97–105
shared property economy 194–5
shelter, State of the World's Cities 64–5
sheltered accommodation 150
Sheppard, S. 82
Shiller, R. 11, 35
short-run adjustment process 34
short-run supply curve 27
signal processing 10
simple model of the economy 16–21
single agency contracts 110
smart cities 214
smart city initiatives 212–13
smart contracting of property sales 196
smart metering 202
social exclusion, urban areas 204
Social Homebuy 102
social migration 119
social mobility, urban areas 203–4
social rental housing 104–6
society, State of the World's Cities 65
socio-political influences, housing markets 6–7
soft human factors 9–10
Souleles, N. 11
sources for housing data 37–8
sources of mortgage financing 173–8
spatial distribution of migrants 132, 134
spatial Durbin model 125
spatial fixity, housing markets 4
spatial hedonic framework 50
spatial planning 76, 79
spatiality, housing markets 4
specialised accommodations 150
stair-casing 98
standard life-cycle housing model 88
State of the World's Cities, UNCHS (United Nations Centre for Human Settlements) 64–6
student housing 183–6
sub-regional housing market areas 36–7
supply 27; aggregate supply function 33; student housing 184–6
supply curve 27–8
supply of intermediate inputs, location choice 70
sustainability, ranking of cities 206–11

symbiotic business model (SBM) 216–17
System of National Accounts 18

tables: Affordability across England and Wales 97; Average number of rooms per person in Europe 159; Basic overview of intermediate housing (SO) schemes in England 102; Definition of smart city 214; Demand drivers in real estate sectors 36; Dwelling stock: By tenure, global comparison 91; Dwelling stock in urban and rural areas 64; The European digital city index 2016 210; Example of hedonic estimation results 49; Example of hedonic estimation results: Neighbourhood effects 83; Forecasts of household numbers 65 and above by age cohort and housing tenure 156; Global Connectivity Index 72; The global Liveability index 2018 210; Global real estate transparency index 2018 40; Government support to mortgage market 174; Growth rate of loans for house purchase 175; Guidelines for the user cost method 89–90; Household tenure status, England and Wales Census 2011 94; Housing deprived population across the income distribution 99; Housing loan penetration by country (2011) 165; Housing's contribution to GDP performance (US) 23; Income/employment impacts of residential construction on the US economy 23; International real estate brokerage commission rate comparisons 112; Knowledge Economy Index (2012) 71; Largest cities in terms of population 2018 200; Life expectancy at age 60 across the world 154; Life expectancy at age 65 (years): United Kingdom 155; Overcrowding rates in households across the income distribution 160; Real estate agency definitions in England 110; Summary of social rental housing across selected countries: Criteria assessed in selecting eligible households 104–5; Sustainable cities 2016 207;
Ten evaluation aspects of homeowners/householders in relation to housing choice 14; Top 100 sustainable cities 2016: People 208; Top 100 sustainable cities 2016: Profit 209; Top 100 sustainable cities 2016: Planet 208; Total and social rental housing stock across selected countries 106; Total housing stock in OECD and EU countries, selected years 29; Total outstanding residential loans to GDP ratio, 2004-2014 166; Twenty highest top marginal corporate tax rates in the world 70; UK house price indices 53; US house price indices: A comparison 54; US recession dates: Business cycle reference dates 22; Variable-rate loan characteristics 177

taxes: location choice 70, 83–4; Pigouvian tax 78
technology application areas 193–7
technology for residential real estate: digital housing 220–1; investments 188–90
tenure choices 88–97
tenure neutrality 87
tenure-mix 98
territoriality, housing markets 4
terrorism, urban areas 205–6
theory of the land rent 73
theory of urban land 73
Thunen, V. 73
Tiebout, C. 80–1
Tiebout model 78–81
Tiwari, P. 45
traffic congestion, urban areas 202–3
transaction costs, housing markets 6
transparency, Global Real Estate Transparency Index 39–40
transport, urban areas 202–3
transport costs, location choice 66
triad of semiotics 12–13
Tuchener, J. 123
types of mortgage financing 168–71

Urban Operating System (UOS) 219
ubiquitous cities 214
U-City 214
UIDAI (Unique Identification Authority of India) 200
unaffordable housing, urban areas 203
unbalanced budgets, urban areas 205
UNCHS (United Nations Centre for Human Settlements), facts and figures 64–6
Unique Identification Authority of India (UIDAI) 200
United Kingdom: affordable housing 135–9; ageing populations 152; ARMs 176; Barker Review 77; EPC (energy performance certificate) 201; GDP (Gross Domestic Product) 18; house price indices 53–7; housing contribution to GDP 23; housing for young people 141–5; immigration 120; life expectancy 154–5; migration 129–31; PRS (private renting sector) 180; spatial distribution of migrants 134; tenure choices 92–6
United Nations Centre for Human Settlements (UNCHS), facts and figures 64–6
United States: brokerage industry 110; house price indices 53–7; housing contribution to GDP 21–3; immigration 120, 125–6; migration 131–2; mortgage financing 171; real estate brokers 110–11; sources of mortgage financing 174; tenure choices 91
UOS (Urban Operating System) 219
urban agglomerations 63–4

urban areas 198–9; challenges of 199–206; models of 72–6
urban big data 218–19
urban pollution 201–2
urban resilience, lack of 204–5
urbanisation 7; global urbanisation 61–2; land-use planning 76–80; location choice 66–72; megacities 63–4; models of urban areas 72–6; regional comparisons 62–3
user cost of housing 85–6

variability, housing markets 5
variable rate mortgages 169, 177
Veblen, T. 7–8
Veblen goods 26
violating law of demand 26–34

virtual cities 214
Von Thunen model 73
Vuchelen, J. 11

Wachter, S. 124
Wallace, A. 101
water insecurity, urban areas 201
White, M. 114
wired cities 214

Yavas, A. 113, 115
Yeh, J. 36–7, 50, 59
Yinger, J. 80
young people, housing for 141–5

zoning 76